Creativity and Innovation for a Better World

Edited by Diana Dias
and Claisy Maria Marinho-Araujo

Published in London, United Kingdom

Creativity and Innovation for a Better World
http://dx.doi.org/10.5772/intechopen.108562
Edited by Diana Dias and Claisy Maria Marinho-Araujo

Contributors
Kenneth Y. T. Lim, Nguyen Duc Minh Anh, Nguyen Thien Minh Tuan, Ahmed H. Hilmy, Josam Nandwa Nandwa Musambayi, Mashitah Binti Sulaiman, Mohd Nazir Ahmad, Ruey-Yun Horng, Ching-Wen Wang, Yun-Chieh Yen, Ting-Yu Wu, Enriko Ceko, Bronislaw Czarnocha, William Baker, Bengt Köping Olsson, Diana Da Silva Dias, Maria Teresa Ribeiro Candeias, Cristiano Pedroso-Roussado, Antonio Cruz Rodrigues, Joao Cunha, Ana Jorge

First published in London, United Kingdom, 2023 by IntechOpen
IntechOpen is the global imprint of INTECHOPEN LIMITED, registered in England and Wales, registration number: 11086078, 5 Princes Gate Court, London, SW7 2QJ, United Kingdom

British Library Cataloguing-in-Publication Data
A catalogue record for this book is available from the British Library

Additional hard and PDF copies can be obtained from orders@intechopen.com

Creativity and Innovation for a Better World
Edited by Diana Dias and Claisy Maria Marinho-Araujo
p. cm.
Print ISBN 978-1-83768-867-8
Online ISBN 978-1-83768-868-5
eBook (PDF) ISBN 978-1-83768-869-2

For EU product safety concerns:
IN TECH d.o.o., Prolaz Marije Krucifikse Kozulić 3, 51000 Rijeka, Croatia,
info@intechopen.com or visit our website at intechopen.com.

Meet the editors

Diana Da Silva Dias is a psychologist with a master's degree in Psychology. She also holds a Ph.D. in Educational Sciences and completed public aggregation exams in psychology and management in 2017 and 2022, respectively. She chaired the Executive Committee of two Ph.D.s, one in management and another in psychology. She was vice-rector for research, quality and academic innovation. A full professor, Dr. Dias is currently a pro-rector for research and innovation. She is also the dean of two faculties of Economic, Social and Business Sciences (Lusófona University Porto and Lusófona University Lisbon). She is a senior researcher at the Center for Research on Higher Education Policies (CIPES). She has published extensively in indexed scientific journals and is the author of several books on management, psychology, and education. She is also a consultant on higher education policies at the European University Association (EUA), the United Nations Educational, Scientific and Cultural Organization (UNESCO), and the Organisation for Economic Co-operation and Development (OECD).

Claisy Maria Marinho-Araujo is a psychologist with master's and doctoral degrees in Psychology from the University of Brasília (UnB), Brazil, three post-docs and two senior internships at the University of Minho, Portugal. She is a professor and researcher at the Institute of Psychology and the Postgraduate Program in Developmental and School Psychology, UnB. She is a consultant for higher education at Instituto Nacional de Estudos e Pesquisas Educacionais Anísio Teixeira (Inep), at the Brazilian Education Ministery (MEC). She is a member of the Technical Advisory Committee of the National Student Performance Exam (ENADE); evaluator at Coordenação de Aperfeiçoamento de Pessoal de Nível Superior (CAPES); coordinator of the School Psychology Laboratory, Psychology Institute, UnB; coordinator of the School Psychology Working Group (ANPEPP); and coordinator of the graduate program in developmental and school psychology, UnB. Dr. Marinho-Araujo was included in the 2021 Alper-Doger Scientific Index of the top 10,000 scientists in Latin America.

Contents

Preface

More than ever, today's world needs creative and innovative ideas to deal with the increasingly demanding challenges it must face. Only new ideas and their practical implementation can address the world's contemporary constraints. In this emerging era, creativity and innovation are key pillars to drive growth, face challenges, and stand out in a competitive world. This is why it is so urgent to reflect on how creativity and innovation can be fostered in communities, especially among the younger generation who will be the protagonists of inevitable change in the future. The widespread recognition of the vital importance of both creativity and innovation in building a more balanced world between environmental, corporate, social, cultural, and personal demands becomes critical.

Despite this link between creativity and innovation, they are not one and the same. Creativity refers to the process of generating new and original ideas. Innovation, on the other hand, involves the fulfillment of creativity into something useful, relevant, and distinctive.

Creativity is an intrinsic ability of every human being. In a creative way, the human brain can conceive ideas and develop thoughts that can go beyond the standards. In other words, creativity exists within our minds and can be routed into action.

Innovation, on the other hand, is more linked to action. Innovation implies creating or proposing changes in systems that are seen as stable. Innovation means offering a new action, function, skill, or improvement to a system already known, making significant improvements to it. Innovation is about creating something new that meets a real need, positively impacting quality of life and human development.

Creativity comes before innovation. Creativity is the moment when references are gathered, and a product or service is idealized and even designed. However, if it is not put into practice, it is just a creative result.

Innovation happens when the idea is put into practice and generates value. In fact, that's why creativity is not measurable, but innovation is. Although creativity and innovation are in different arenas, they are forces that can coexist and, when they intertwine, yield improved outcomes.

This book focuses on the core and profitable link between creativity and innovation. It contains a set of reflections, experiences, and thoughts on how these two concepts become positive catalysts for each other. It also focuses on how to promote creativity and innovation in people, organizations, and communities.

This book is organized into three sections: "To Drive Creativity and Innovation Forward", "Creativity and Innovation: An Eternal Bond", and "Other Worlds for Creativity and Innovation".

Section 1, "To Drive Creativity and Innovation Forward", focuses on the importance of promoting and reinforcing the importance of developing creative skills. It includes three chapters.

Chapter 1 reflects on the importance that higher education has given to the teaching/ development of creativity skills among its graduates. It addresses how creativity is defined in the scope of learning outcomes of higher education. Results show that some scientific fields, such as engineering, information and media, humanities, life sciences, agriculture, forestry and fisheries, and arts and architecture tend to value more creativity as a desirable learning outcome. Conversely, law, education, social sciences, and business sciences are explicit examples of the undervaluation of competencies related to creativity as part of the professional profile of their graduates. Findings help to understand the real importance that higher education institutions attribute to creativity in their academic curricula.

Chapter 2 is based on the author's research on group creativity and educational initiatives in higher education courses at an advanced level. It highlights correlations between experiences of flow in idea-generating group activities and group members' integrative social behavior. In addition, the chapter shows that training for increased divergent thinking also develops broadened attention, openness, and flexibility for perspective shifts. The chapter presents a conceptual framework to construct a model of research design on collaborative creativity with the purpose of enabling comparisons between the study's methodology and findings to continue developing this field of research with joint efforts. The chapter advances the view that the impact of activities training group-based creativity should be raised in significance beyond individual brilliant ideas because creative collaboration develops abilities to take initiative, make decisions and interact constructively together.

Chapter 3 presents a new theory of creativity in mathematics education. The chapter explores the Koestler/Prabhu theory of learning through Aha! Moment and applies the bisociative frame to investigate its processes of interaction with several different theories of learning. Uncovered processes of interaction suggest that creativity should be the basis of contemporary learning and teaching.

Section 2, "Creativity and Innovation: An Eternal Bond", includes four chapters, all of which contribute to the idea of inseparability between creativity and innovation.

Chapter 4 looks at what tends to hinder creativity in organizations and how creativity informs entrepreneurship. Considering that innovation, entrepreneurship, and creativity are at the heart of entrepreneurship, the chapter focuses on the solutions to bond the creativity gap by refuting its possible barriers. It is a work that aims to mobilize policymakers, academics, and entrepreneurs to effectively use creativity to support innovation and entrepreneurship development.

Chapter 5 presents an integrated model of cognitive tasks involved in the process of technological innovation. The main topic of this model is the proposition that three cognitive conditions are necessary for technological imagination and innovation, namely, (1) cross-domain knowledge, (2) simple heuristics, and (3) pattern recognition ability. Although the required domain knowledge and implementation methods are different across domains, this chapter advocates that heuristics that lead to a breakthrough at each phase in technological innovation are similar, with the conceptual combination as the cognitive engine for generating original and imaginative ideas.

Chapter 6 discusses the importance of creativity and innovation in the pursuit of fashion among women entrepreneurs of the middle class. It examines the factors that led to the involvement of women entrepreneurs in the fashion industry, namely, religious responsibility, self-interest in entrepreneurship, family support, positive attitude towards success, and improving soft skills for survival. The findings indicate the importance of creativity and innovation in the production of Muslim fashion and design products.

Chapter 7 aims to establish the significant connections between innovation and creativity, innovation and the culture of quality management, and creativity and the culture of quality management globally as a response to the crisis and post-crisis period. For the authors, since issues related to innovation, creativity, quality, and quality management have recently attracted more attention than other topics globally, considering psychological factors was of great importance, too. The main finding of this study is that there are relationships between innovation and creativity, between innovation and application of ISO standards, and between quality management and creativity. Hence, quality management, which reflects and refers to the notion of exceeding customer expectations by enhancing the quality of goods and services, is correlated with culture, innovation, and creativity as a response to the crisis and post-crisis period.

The final section of this book, "Other Worlds for Creativity and Innovation", showcases concrete approaches to the practical application of creativity and its enhancement. Chapter 8 presents the "Life^{2}Well Project: Learning at the Intersection of AI, Physiology, EEG, Our Environment and Well-being." In this research project, creativity and innovation are implemented via identical units of a wearable device designed and worn by adolescents that contain environmental sensors that detect ambient temperature, air pressure, infrared radiation, and relative humidity. Results show that the most influential microclimatic factors on biometric indicators were noise and the concentrations of carbon dioxide and dust. Such findings suggest implications for the design of living conditions with respect to the interaction of microclimate and human health and comfort.

Finally, Chapter 9 highlights the contribution of design as a discipline to the reflection of creativity enhancement. It presents the Biology of Creativity Model (BoC), which is a design-by-analogy method that promotes an empowered design and creative practice through analysis of mostly biology reference texts for enhanced creative performance in a diverse range of contexts.

From different cultures, different scientific fields, different research methodologies, the different contributions that make up this book shape an integrated approach to creativity and innovation as central concepts to contribute to a better society.

We believe that creativity and innovation are indeed the forces that drive progress, expression, and discovery. Investing in them is investing in a better, more balanced, and sustainable world.

Diana Dias
Lusofona University,
Lisboa, Portugal

Claisy Maria Marinho-Araujo
University of Brasilia,
Brasilia, Brasil

Section 1

To Drive Creativity and Innovation Forward

Chapter 1

Can Creativity Be Taught and/or Learned? A Sketch from Higher Education Learning Outcomes

Diana Dias and Maria Teresa Ribeiro Candeias

Abstract

Creativity embodies the ability to discover new and innovative ideas, links, and problems' solutions. Creativity is a competitive advantage that enables both individuals and organizations to succeed in increasingly demanding markets. Therefore, the development of this competence has become central to the debate on curricula within education systems and the labor market. So, the present study seeks to address how creativity has been defined in the scope of learning outcomes of the new study programs under quality evaluation and subsequent accreditation in the Portuguese higher education system. Adopting a qualitative approach, a content analysis of these learning outcomes was carried out using a theoretical and empirically validated conceptual matrix. Results point out for universities and the public sector of Portuguese higher education system tend to value more creativity as a desirable learning outcome, as well as some scientific fields, such as Engineering, Information and Media, Humanities, Life Sciences, Agriculture, Forestry and Fisheries and Arts and Architecture. Conversely, Law, Education, Social Sciences and Business Sciences are explicit examples of the undervaluation of competences related to creativity as part of the professional profile of their graduates. Findings help to understand the real importance that Portuguese higher education institutions attribute to creativity in their academic curricula.

Keywords: creativity, higher education, curriculum, learning outcomes, skills

1. Introduction

In a global and highly competitive context, the issue of creativity is particularly relevant. Indeed, creativity is a competitive advantage that enables both individuals and organizations to succeed in increasingly demanding markets. Therefore, the development of this competence has become central to the debate on curricula within education systems and the labor market [1]. Several authors stress the significance of promoting creativity skills in areas such as the design of new services, new technologies and production processes within organizations or in areas such as health and education [2, 3]. Montuori and Donnelly [4] identify several fields of knowledge that deal with creativity, such as psychology, sociology, but also marketing, art, or anthropology. However, the influence of creativity is not limited to the organizational dimension. At the individual level, creativity is a form of self-expression and a way of promoting engagement and a sense of accomplishment. Individual creativity is seen as a uniquely human trait that generates personal satisfaction and positive emotions [5]. Literature review suggests that creativity has an impact on individuals' quality of life, as it generates feelings of satisfaction and pleasure, which are essential for mental health and emotional well-being [6–9].

It is therefore relevant to realize the concept of creativity. Like other concepts, creativity is approached and consequently defined from different theoretical perspectives, which, when combined, allow for a more robust understanding of the concept, even if they are contested [10].

According to Sarsani [11], there is general agreement that the concept of creativity should be approached in a comprehensive manner, considering its multiple dimensions. The first allusions to the concept of creativity [12] focus on the abilities revealed by creative people. Various authors converge in defining creativity as the production of new ideas with utility or as a problem-solving proposal [13, 14].

The psychological concept of creativity identifies two components: originality, which refers to novelty and uniqueness, and usefulness, which entails suitability and adaptation to a group or culture [15, 16]. In a similar exercise, Rhodes [17] proposes a structure consisting of four components: person, product, process, and press, which are interrelated and overlapping. Thus, person reflects the characteristics that reveal creative potential; process refers to the cognitive and affective processes that drive creativity; product reflects the characteristics of the creative outcome; and press alludes to the effect of the environment and outside influences as stimulators or blockers of creativity. Amabile and Pillemer [18] understand creativity as a socio-psychological phenomenon. On an individual level, Lindqvist [19] argues that creativity is a construct of each individual imagination. However, the individual characteristics of creators are related to their culture and environment [20]. This interconnection with culture and environment reflects that, despite the individual nature of the act of creation, it takes place within a system of social relations [21]. The results of Kampylis and Valtanen [22] work show that the different definitions of creativity intersect in four elements. First, creativity is characterized as an individual talent. It also implies a conscious process that takes place in a specific environment and results in a tangible or intangible output [22]. While listing the various proposed components, Walia [23] defines creativity as an action that arises from the perception of the context, which identifies a certain imbalance that gives rise to a productive activity capable of challenging norms and thought patterns, resulting in something new that appears as a material object or a mental or emotional structure.

Over time, the limits of human potential in several fields have often been redefined by accidental conditions and creative insights. These moments of creativity have stimulated progress, reshaped our lives, and transformed the understanding of individuals, pushing the boundaries of what was once unknown as possible or not [24–26]. While creativity is influenced by the creative domain and field, individual creativity potential refers to the creative actions that individuals can perform in their everyday lives. All individuals possess this potential, as it corresponds to their ability to construct original interpretations and is identified as a non-specialized creative thinking process that does not depend on a specific domain or field. However, creative thinking is a crucial competence for all learners and therefore for education. Although creativity and innovative thinking are fundamental components of educational technology, their absence in educational curricula is a serious limitation of modern education in all academic disciplines [5, 27–32].

In fact, creativity becomes increasingly significant for the professional, personal, and social profile of all higher education graduates in all scientific areas. If in the Renaissance, creativity might be a luxury available to only a few, at present, it is an unavoidability for the survival off all [33]. In fact, creativity is assumed as a competence that supports not only the ability to clutch opportunities, but also to cope effectively with challenges and hitches in the personal, social, and professional fields. The advantages of creativity are recognized in the world of work, as it is considered a key competence for innovation and organizational success [3].

Sustainability is more and more linked to each country's capability to innovate, create new products and services, develop new technologies and production methods, offer products and services to new markets, and, in the global context, address the wide range of challenges in the fields of health, education and labor. [2]. Recommendations to uplifting creativity in higher education curricula from China and Japan governments, as well as by American and European business top managers [34]. In line, Lubart [35] claims that, "in respect to the capital issues, such as those of the social or planetarium equilibrium, the need of new approaches and solutions becomes increasingly urgent" (p. 8). And if creativity is accepted as a fundamental competence for success in the labor market, it should necessarily be a skill to be contemplated in the educational curriculum, with special emphasis on higher education. Particularly for Higher Education, boosting knowledge, innovation, and creativity emerges as an intentional undertaking, appealing for creative research, creative pedagogies, and creative organizational structures. Jackson [36] stresses that higher education must recognize the critical role of creativity in training youth for the unpredictable and challenging labor market. He argues that promoting creativity-related skills would be part of their higher education experience, promoting their potential as far as possible. Several researchers argue that creativity plays a central role in the knowledge society and that higher education should actively promote strategies, methods, procedures, and guidelines for the development of students' creativity [36–39]. As stressed by European University Association [40], for creativity to thrive in higher education, higher education institutions and external stakeholders need to promote purposeful and determined efforts to do so. This organization developed the Creativity in Higher Education project, involving 42 higher education institutions based in 21 different countries. This project aimed to understand and explore not only the factors promoting, but also the factors inhibiting the emergence of creativity in the context of higher education. Its approach focused on innovation in teaching and learning and the structure and leadership of higher education institutions [40]. Project findings provide strong endorsement to the idea that diversity and combination of arts and other disciplines can provide a highly favorable environment for addressing a multilayer issue such as creativity. In fact, diversity was recognized as a critical driver for strengthening creativity, whether applied to research teams, to teaching and learning methods, or to joint projects with external partners. However, scientific evidence suggests that creativity decreases with increasing years of formal education [33, 41–44]. According to these authors, dissuasion and penalization of creativity expression in higher education seem to be the reasons behind this issue. In fact, Cropley [45] presents a clear example of creativity relegation by higher education institutions. The author found that Australian universities do not provide the necessary guidance for students to master successful strategies to cope with new situations, to cope with rapid change and to address changing failures, as 75% of all recent alumni, regardless of their scientific field, were recognized by employers as lacking creativity, problem-solving and critical and independent thinking skills. In the same vein, Jackson et al. [38] argue that the importance of creativity in the teaching and learning process has been rather undervalued, as higher education institutions tend to value skills such as critical thinking more than creativity. Also in Brazil, Castanho [42] detected the diminished importance attached to creativity in higher education curricula. In the same vein, Hosseini [46] draws attention to the faculty's difficulty in relying on educational practices that intentionally foster students' motivation and creativity.

Jackson [36] also points to the attitudes and resistance of not only faculty staff but also students themselves to pedagogical methodologies that explicitly foster creativity.

The same author identifies other potential institutional barriers to fostering creativity in higher education, such as structural, cultural and procedural factors, lack of time and other resources, or institutional policies lacking creativity promotion. Despite the recognition that the development of creative skills is an asset to graduates' personal, social, and professional profiles, creativity is very seldom embedded in higher education curricula as an overt learning outcome.

This chapter aims specifically to explore creativity as an explicit learning outcome in higher education curricula in Portugal. In other words, it aims to understand the weight of creativity in the graduate profile that higher education institutions claim to develop.

2. Methodology

In order to analyze the content of the learning outcomes mentioned by the Portuguese higher education institutions for their courses, the documents that these institutions may have presented to the Assessment and Accreditation Agency for Higher Education (A3ES) were considered. In the initial phase of the accreditation process, the higher education institutions submit to A3ES an accreditation proposal: "Previous Accreditation Request of a New Study Cycle." For each assessed program, the learning outcomes that students are expected to achieve on graduation are listed. By an internal rule of A3ES, the information provided in this context is limited to 1000 characters.

Thus, 2.926 evaluation and accreditation request processes were analyzed for New Study Cycles and Study Cycles already in operation.

A content analysis of the learning outcomes acknowledged in the proposals of study programs proposed to quality accreditation by the Portuguese Agency for Higher Education Accreditation (A3ES) was carried out using the MAXQDA software (version 12). Six steps were taken to identify patterns (themes) within the data. This analysis followed a conceptual framework, in which 24 technical and generic skills were included. This procedure uses a theme-based analysis approach, rather than a data-driven approach [14]. Creativity was one of the categories found and analyzed. Within the category Creativity, 6 sub-categories were considered: to Create, To be Original, To solve new problems, to go beyond, to think out of the box, and To transform. Learning outcomes examples for each sub-category could be respectively:

- "Ability to create, undertake and innovate";
- "With the conclusion of the program it is anticipated that the student will be able to solve problems in an original way, alone or in collaboration";

- "Training of postgraduates prepared to solve new problems with creativity and sensitivity to changes taking place in the present world";
- "Ability to problematize the dynamics inherent to historical processes in a critical, creative and independent way, and capable of suggesting valid answers, scientifically based";
- "Ability to innovate, ability to formulate unconventional questions and to think of alternatives"
- "The mission of the program is to train competent and enterprising professionals with the ability to use technology (software and hardware) in a creative way, innovating and adding value to the national ICT business fabric (new applications, products, services)."

As already mentioned, the data analysis was focused on the information included in the documents submitted for quality accreditation and that refers to the "intended learning outcomes" that students are expected to achieve at the end of a given learning period. All learning outcomes (n = 2926) included in all proposals of new study cycles submitted to the A3ES for accreditation were analyzed. Of these 619 documents, 54.8% were 2nd study cycle proposals, while 26.2% referred to the 1st study cycle and 18.9% to the 3rd study cycle.

The content analysis of all learning outcomes in isolated categories, that is, according to the conceptual synthesis matrix, made it possible to obtain a remarkable set of descriptive data, both by competence and by independent variable. These data thus made it possible to understand which competences and knowledge are most valued not only by Portuguese Higher Education in general but by each subsystem, sector, scientific area, etc. To this end, the quantitative data resulting from the previous content analysis were explored using the statistical program IBM SPSS for Windows, version 25 (IBM Corp. Released, 2010).

First, the univariate normality of all variables was confirmed using the Kolmogorov-Smirnov test and the asymmetry (values <1.0) and kurtosis (values <3.0) criteria, as defined by Kline [47], and the non-existence of outliers ($|z| < 3$; [47]). Whenever any of the assumptions were not checked, the corresponding non-parametric tests were performed. When both tests were concordant regarding the rejection versus retention of the null hypothesis, the parametric tests were reported [48].

In the analysis of the results, the statistical procedures used included not only descriptive statistics but also parametric tests for independent samples (Student's t-test and one-factor analysis of variance) and relationships between variables through Pearson's r coefficient.

It should be noted that, in the ANOVAs, after the homogeneity of the variances of the variables used was tested through Levene's test, the Gabriel post hoc test was used given the unequal number of subjects in each of the groups studied [49].

3. Results

Results point out the significance ascribed to Creativity as a learning outcome. To do that, the frequency of mentions of creativity in different higher education courses in Portugal was explored.

When it has analyzed the weighting of the competences that appear as expected in the professional profile of each graduate, critical thinking is the most common learning outcome in 74.3% of the study cycles analyzed. Thus, critical thinking could be seen as the main transversal competence that the Portuguese academia intends to develop in its students. Most of the higher education offers (62.7%) tend to also value the competence to solve problem, as well as personal development (52.5%) and information management (50.1%). But, if 49.1% of all study programs mention leadership and 47.1% refer to ethics as a learning outcome expected at the end of the degree, only 46.2% raise creativity as a skill to develop. With smaller percentages, we find communication and entrepreneurship which are referred to by 46.2% and 41.2% of the courses analyzed. **Figure 1** shows the percentage of mention of each skill in all study programs.

As can be observed in **Table 1**, the Gabriel post hoc test reveals that creativity presented a lower relevance in undergraduate degrees, followed by masters' degrees. The relevance assigned by the undergraduates to this learning outcome differed significantly from the importance assigned to it in all other degrees. The same was true for masters' degrees, where significant differences were found in comparison with the other degrees. Only in the integrated masters and doctoral degrees were there no significant differences between the importance attributed to creativity as a learning outcome.

Learning outcomes associated with creativity seem to be more valued by the university subsystem, and these differences are statistically significant.

On the other hand, learning outcomes associated with creativity seem to be more valued by the university subsystem and the public sector, and these differences are statistically significant. These differences were not found regarding seniority and geographical location of the higher education institutions integrating the analyzed courses.

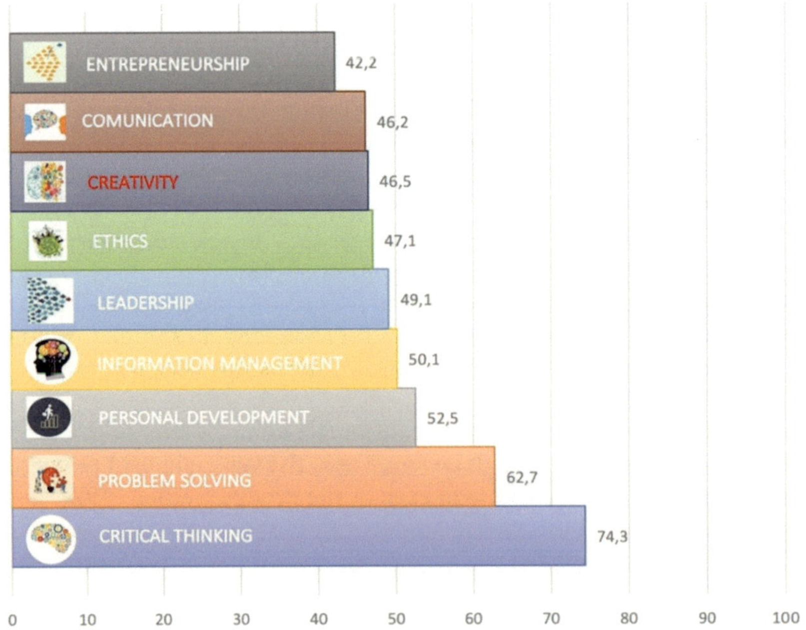

Figure 1.
Percentage of mention of each skill in all study programs.

	Bachelor	Master	Integrated Master	PhD	F		η_p^2
	A (SD)	A (SD)	A (SD)	A (SD)	GL	Value	
Creativity	0.61 (0.94)	0.77 (1.20)	1.31 (1.47)	1.30 (1.30)	(32922)	40.92***	0.04

*Note. A = median; SD = standard deviation. ***p < .0001. According to Cohen (1988): ηp2]0.00–0.05]—small effect size.*

Table 1.
Distribution of learning outcomes by academic degree.

When analyzing the importance attributed by each scientific area to creativity as a learning outcome, interesting trends can be observed. Thus, creativity is mentioned for 30.3% of the degrees in Education, 52.2% in Arts and Architecture, 60% in Humanities, 35% in Social Sciences, 61.6% in Information and Media, 35.4% in Business Sciences, 33.3% in Law, 55.6% in Sciences, 68.6% in Engineering, 56.5% in Agriculture, Forestry and Fisheries, 21.7% in Health Sciences, and 46% in Services.

The Engineering cycles are the ones that most promote innovation and creativity as competences to be developed by their students, significantly different from Education, Arts and Architecture, Humanities, Social Sciences, Information and Journalism, Business Sciences, Law, and Agriculture. On the other hand, it is the courses in the Health area that least report innovation and creativity as an expected learning outcome. In terms of significant differences among the various scientific areas regarding the weight given to creativity in their study cycles are (i) Education in relation to Arts and Architecture, Humanities, Information and Journalism, Sciences, Agriculture and Services, (ii) Arts and Architecture in relation to Social Sciences, Business Sciences, Law and Health; (iii) the Humanities and Information and Journalism, both in relation to Business Sciences and Health; (iv) Business Sciences in relation to Agriculture and Services; (v) Law in relation to Sciences; and (vi) Sciences in relation to Agriculture and Health.

4. Conclusions

For Portuguese academia, creativity could be seen as the skill to create new things or new ideas. Creativity and originality are typically used as synonymous. But could also be noticed to solve new or renew problems. Creativity could be as well the way to go beyond, think out of the box, or to transform problems and to create new answers.

In fact, creativity is a learning outcome cited for only 46.5% of study programs proposed to quality accreditation by the national agency for quality assurance of higher education in Portugal. However, when the focus is put on the frequency of each category coded, creativity is the third learning outcome least cited in the Portuguese academy. If according to Jackson et al. [38], higher education institutions tend to give more importance to critical thinking, but our results give up the idea of the significance of creativity in teaching and learning processes have been widely undervalued in higher education.

Portuguese academia clearly underestimates the importance of fostering creative skills in its bachelor and master students, which clearly runs counter to trends in the world of work and policy guidelines.

Universities (compared to more polytechnic institutions) and the public sector of Portuguese higher education system tend to value more creativity as a desirable learning outcome.

Most of the degrees in Engineering, Information and Media, Humanities, Life Sciences, Agriculture, Forestry and Fisheries, and Arts and Architecture mention expressly creativity as a learning outcome. Health Sciences is the scientific area that least values creativity, since less than 22% of its courses refer to creativity as one of its learning outcomes. But if these results can be, at least in part, justified by the fact that the health sciences teaching/learning process is very focused on clinical protocols, as law focused on regulations, the same cannot be justified in the case of other scientific areas. Education, Social Sciences, and Business Sciences are explicit examples of the undervaluation of competences related to creativity as part of the professional profile of their graduates.

Advocating Wisdom [50], Portuguese higher education system seems to need a cultural change, "to help teachers understand and enhance their own creativity and to recognize this as an integral part of their professionalism" (p. 183), as well as warranting an institutional climate that boosts personal development not only of the students but also of the faculty staff.

To fostering creativity in higher education, it is critical the intentional development of an institutional culture that enhances creativity and its expression. Boosting active learning and inspiring students to be creative, originals and innovative is one of the ways of assuring that creativity was in fact a real learning outcome of higher education.

Acknowledgements

We would like to thank Agency for Assessment and Accreditation of Higher Education—A3ES for providing the data that allowed us to develop this research.

Conflict of interest

The authors declare no conflict of interest.

Author details

Diana Dias[1]* and Maria Teresa Ribeiro Candeias[2]

1 Lusofona University, Lisboa, Portugal

2 Lusofona University, Porto, Portugal

*Address all correspondence to: diana.dias@ulusofona.pt

References

[1] Dias D, Soares D. Creativity is intelligence having fun: Is creativity a real learning outcome of higher education? In: Edulearn18 Proceedings. Palma de Mallorca: IATED; 2018. pp. 6124-6129

[2] Smith-Bingham R. Public policy, innovation and the need for creativity. In: Jackson N, Oliver M, Shaw M, Wisdom J, editors. Developing Creativity in Higher Education. London: Routledge; 2007. pp. 10-18

[3] Bruno-Faria M, Vargas E, Martínez A, editors. Criatividade e inovação nas organizações. Desafios para a competitividade [Creativity and innovation in organizations. Challenges for competitiveness]. São Paulo, SP: Atlas; 2013

[4] Montuori A, Donnelly G. The creativity of culture and the culture of creativity research: The promise of integrative transdisciplinarity. In: Glăveanu V, editor. The Palgrave Handbook of Creativity and Culture Research. London: Palgrave Macmillan; 2016. pp. 743-765

[5] Root-Bernstein R, Root-Bernstein M. Sparks of Genius: The Thirteen Thinking Tools of the world's most Creative People. Boston, Mass: Houghton Mifflin Co; 1999

[6] Alencar E. O papel da escola no desenvolvimento do talento criativo [the role of school in the development of the creative talent]. In: Fleith D, Alencar E, editors. Desenvolvimento de Talentos e Altas Habilidades [Developing Talents and High Ability]. Porto Alegre, RS: ArtMed; 2007. pp. 151-162

[7] De Breu C, Baas M, Nijstad B. The emotive roots of creativity: Basic and applied issues on affect and motivation. In: Mumford M, editor. Handbook of Organizational Creativity. San Diego, CA: Elsevier; 2012. pp. 217-240

[8] Pannells T, Claxton A. Happiness, creative ideation, and locus of control. Creativity Research Journal. 2008;**20**(1):67-71

[9] Runco M. Creativity. Annual Review of Psychology. 2004;**55**(1):657-687. DOI: 10.1146/annurev.psych.55.090902.141502

[10] Kozbelt A, Beghetto R, Runco M. Theories of creativity. In: Kaufman J, Sternberg R, editors. The Cambridge Handbook of Creativity. Cambridge: Cambridge University Press; 2010. pp. 20-47

[11] Sarsani M. Creativity: Definition and approaches. In: Sarani M, editor. Creativity in education. Delhi: Sarup & Sons; 2005:1-7

[12] Guilford J. Creativity. American Psychologist. 1950;**5**:444-454. DOI: 10.1037/h0063487

[13] Amabile T, Barsade S, Mueller J, Staw B. Affect and creativity at work. Administrative Science Quarterly. 2005;**50**(3):367-403

[14] Woerkum C, Aarts M, Grip K. Creativity, planning and organizational change. Journal of Organizational Change Management. 2007;**20**(6):847-865

[15] Stein M. Creativity and culture. Journal of Psychology. 1953;**36**:311-322. DOI: 10.1080/00223980.1953.9712897

[16] Barron F. The disposition towards originality. Journal of Abnormal and Social Psychology. 1955;**51**:478-485. DOI: 10.1037/h0048073

[17] Rhodes M. An analysis of creativity. Phi Delta Kappan. 1961;**42**(7):305-310

[18] Amabile T, Pillemer J. Perspectives on the social psychology of creativity. The Journal of Creative Behavior. 2012;**46**:3-15

[19] Lindqvist G. Vygotsky's theory of creativity. Creativity Research Journal. 2003;**15**:245-251

[20] Lebuda I, Csikszentmihalyi M. All you need is love: The importance of partner and family relations to highly creative individuals' well-being and success. The Journal of Creative Behavior. 2018;**54**:1-15

[21] Glăveanu V. Rewriting the language of creativity: The five A's framework. Review of General Psychology. 2013;**17**(1):69-81. DOI: 10.1037/a0029528

[22] Kampylis P, Valtanen J. Redefining creativity—Analyzing definitions, collocations, and consequences. The Journal of Creative Behavior. 2010;**44**:191-214. DOI: 10.1002/j.2162-6057.2010.tb01333.x

[23] Walia C. A dynamic definition of creativity. Creativity Research Journal. 2019;**31**(3):237-247

[24] Songca R. Transdisciplinarity: The dawn of an emerging approach to acquiring knowledge. International Journal of African Renaissance Studies - Multi-, Inter- and Transdisciplinarity. 2006;**1**:221-232. DOI: 10.1080/18186870608529718

[25] Montuori A. Systems approach. In: Runco M, Pritzker S, editors. Encyclopedia of Creativity. 2nd ed. San Diago, CA: Academic Press; 2011. pp. 414-421

[26] Glaveanu V, Hanson M, Baer J, Clapp E, Corazza G, Hennessey B, et al. Advancing creativity theory and research: A socio- cultural manifesto. Journal of Creative Behavior. 2019;**0**(0):1-5. DOI: 10.1002/jocb.395

[27] Csikszentmihalyi M. Creativity: Flow and the Psychology of Discovery and Invention. New York, NY: Harper Collins; 1996

[28] Kaufman J, Beghetto R. Beyond big and little: The four c model of creativity. Review of General Psychology. 2009;**13**(1):1-12. DOI: 10.1037/a0013688

[29] Lewis T. Creativity in technology education: Providing children with glimpses of their inventive potential. International Journal of Technology & Design Education. 2009;**19**(3):255-268. DOI: 10.1007/s10798-008-9051-y

[30] Robinson K. Mind the Gap: The Creative Conundrum. Critical Quarterly. 2001;**43**:41-45

[31] Runco M. Creativity and education. New Horizons in Education. 2008;**56**(1):96-104

[32] Sternberg R. What is the purpose of schooling? In: Ambrose D, Sternberg R, editors. How Dogmatic Beliefs Harm Creativity and Higher Level Thinking. New York and London: Routledge: Taylor and Francis Group; 2012. pp. 207-219

[33] Csikszentmihalyi M. Developing creativity. In: Jackson N, Oliver M, Shaw M, Wisdom J, editors. Developing Creativity in Higher Education. London: Routledge; 2007. pp. 18-20

[34] Strom R, Strom P. Changing the rules: Education for creative thinking. The Journal of Creative Behavior. 2002;**36**(3):183-200. DOI: 10.1002/j.2162-6057.2002.tb01063.x

[35] Lubart T. Psicologia da Criatividade [Psychology of Creativity]. Porto Alegre, RS: ArtMed; 2007

[36] Jackson N. Creativity in higher education. Creating tipping points for cultural change. SCEPTrE Scholarly Paper. 2006;**3**(1):1-25

[37] Alencar E, Fleith D. Criatividade na educação superior: Fatores inibidores [creativity in higher education: Inhibiting factors]. Avaliação. 2010;**15**(2):201-206. DOI: 10.1590/S1414-40772010000200011

[38] Jackson N, Oliver M, Shaw M, Wisdom J, editors. Developing Creativity in Higher Education. London: Routledge; 2007

[39] Romo M. Psicología de la Creatividad [Psychology of Creativity]. Barcelona, España: Paidós; 2012

[40] European University Association. Creativity in Higher Education: Report on the EUA Creativity Project 2006-2007. Brussels: The European University Association; 2007

[41] Alencar E, Fleith D. Criatividade. In: Múltiplas Perspectivas [Creativity. Multiple Perspectives]. Brasília, DF: Editora da Universidade de Brasília; 2009

[42] Castanho M. A criatividade na sala de aula universitária [creativity in the university classroom]. In: Veiga I, Castanho M, editors. Pedagogia universitária. A Aula Em Foco [University Pedagogy. Focusing the Classroom]. Campinas, SP: Papirus; 2000. pp. 75-89

[43] Wechsler SM. A educação criativa: Possibilidades para descobertas [Creative education: Possibilities for discoveries]. In: Castanho S, Castanho M, editors. Temas e Textos Em Metodologias Do Ensino Superior [Themes and Topics in Higher Education Methodologies]. Campinas, SP: Papirus; 2001. pp. 165-170

[44] Wechsler S. Criatividade e desempenho escolar: Uma síntese necessária [creativity and school achievement: A necessary sinthesis]. Linhas Críticas. 2002;**8**(15):179-188

[45] Cropley A. Creativity in Education & Learning. London: Routledge; 2005

[46] Hosseini A. University student's evaluation of creative education in universities and their impact on their learning. Procedia - Social and Behavioral Sciences. 2011;**15**:1806-1812. DOI: 10.1016/j.sbspro.2011.04.007

[47] Kline R. Principles and Practice of Structural Equation Modeling. New York: Guilford Press; 2011

[48] Fife-Schaw C. Levels of measurement. In: Breakwell G, Hammond S, Fife-Schaw C, Smith J, editors. Research Methods in Psychology. 3th ed. Vol. 2006. London: Sage; 2006. pp. 50-63

[49] Field A. Discovering Statistics Using SPSS. 3rd ed. London: Sage Publications Ltd.; 2099

[50] Wisdom J. Developing higher education teachers to teach creatively. In: Jackson N, Oliver M, Shaw M, Wisdom J, editors. Developing Creativity in Higher Education. London: Routledge; 2007. pp. 183-196

Chapter 2

Collaborative Creativity

Bengt Köping Olsson

Abstract

This chapter is based on the author's research on group creativity and educational initiatives in the private and public sectors as well as in higher education courses at an advanced level. The contribution is derived from both qualitative and quantitative methodological approaches to present in-depth knowledge of creative collaboration and competence as well as training of the necessary skills needed to activate both the prerequisites and predictors for creativity. Research shows correlations between experiences of flow in idea-generating group activities and group members' integrative social behavior. In addition, it is shown that training for increased divergent thinking also develops broadened attention, openness, and flexibility for perspective shifts. A conceptual framework is presented to construct a model of research design on collaborative creativity with the purpose of enabling comparisons between study's methodology and findings to continue developing this field of research with joint efforts. The chapter advances the view that the impact of activities training group-based creativity should be elevated in importance beyond individual brilliant ideas because creative collaboration develops abilities to take initiative, make decisions and interact constructively together.

Keywords: collaborative creativity, social interaction, synchronized flow, mindfulness and openness, fluency and flexibility, research design

1. Introduction

Creativity emerges in day-to-day practice in the space between what is and what is to come, it is, therefore, vital to define creativity as the collective realization of ideas in meaningful ways within social practices [1]. Coworkers' collaboration is important for developing workplaces, therefore we need to understand in what ways creativity in social groups can increase exchange and deepen the interaction quality.

With the aim of developing a conceptual framework for research at the group level, the concept of group idea was defined with inspiration from Mary Parker Follett (e.g., circular responses) [2] and Ludwig Fleck (e.g., thought collective) [3]. In studies on creative processes in social groups' interaction, exchange of ideas, and intersubjective knowledge development, the concept of group idea is central to understanding how emergent content and structures can covary for creativity productivity and performance. This chapter thus scrutinizes the function of and consequences for creativity in social interaction—the exchange between group members' different opinions and perspectives. This, in turn, contributes to better conditions and enable continuous development of knowledge and competencies in relation to both group and individual levels.

Conceptual frameworks for research on social groups' creative processes are often characterized as group dynamics [4–6]. What is denoted dynamics is the ongoing alteration between divergence and convergence as well as the thought ideal of iterations between individual level (group members) and the group level cohesiveness supporting the development of group ideas. However, these kinds of frameworks or models do not describe and explain what constitutes the collaborative situation/group-level state that makes the activities described in these models possible. That is, using the concept of "group dynamics" tends to establish another black box hiding what specific factors of social interaction enable collaborative creativity, for example, emergent outcomes characterized as original.

Instead, researchers (as well as practitioners!) should pay serious attention to creativity research where broadened thinking supports integration of different ideas and perspectives, as well as to social interaction, that is, intensity of dialog and quantity of exchange as drivers of self-organization and development of group maturity. After all, the purpose of developing divergent thinking may not primarily have the goal of efficient individual idea generation. What should be regarded as higher value and relevance of divergent thinking capabilities, is the activation of the so-called executive functions in relation to self-efficacy on an individual level and self-organization on group level.

Expressions such as 'teams are made, not born' are based on the belief that social interaction, dialog, exchange of views, and shifts in perspective can be trained and developed [7], that is, constructive integration of divergent propositions does not just happen "by itself". In other words, groups' collaborative creativity can be strategically trained and given appropriate conditions (i.e., prerequisites). From the perspective of facilitation of creative collaboration, there are two prominent categories of interrelated prerequisites [8]: personal and behavioral characteristics. The level of these prerequisites should be considered in relation to each other as this correlation should guide the direction and scope of training.

Personal characteristics may for ethical reasons not be affected other than by the individual's initiative, for example, motivation to belong and contribute to the group's development. Behavioral characteristics such as attitudes, increased awareness, as well as aptitude, can be influenced through training and knowledge development [4].

In this chapter, the line of reasoning is structured in two main steps (1) that creativity enables social interaction and exchange, and (2) that social interaction drives creativity. **Figure 1** visualizes the potentially mutually reinforcing, iterative correlation between, interaction and creativity, conditioned by certain prerequisites that can be assessed and as well as other types of factors that can be trained and thus predict creativity.

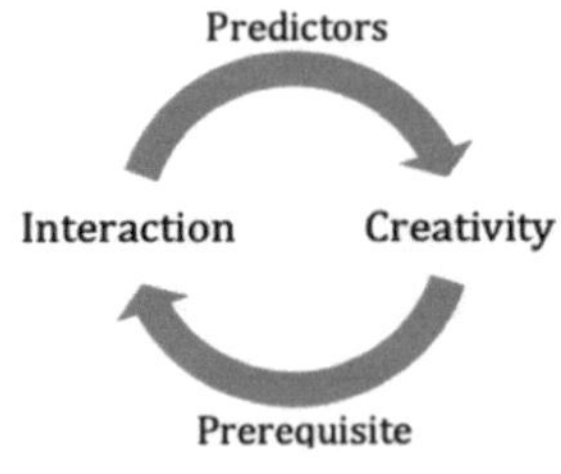

Figure 1.
Visualization of the iterative reinforcing correlation between interaction and creativity conditioned by certain prerequisites and predictors.

This chapter thus investigates how social interaction and exchange can be conceptualized in terms of creativity and what factors (predictors) in social interaction support creativity as well as what factors (prerequisites) emerge in creative processes that enable social interaction and establish the quality of creativity.

Formulated as research questions: *How can social interaction and exchange be understood in terms of creativity? By which research design can predictors and prerequisites for collaborative creativity be related, assessed, and analyzed?*

2. Collaborative creativity and interaction quality

Sternberg, Lubart, Kaufman, and Pretz [9] emphasized that creativity is important for society, but points out that research on problem-solving, abductive reasoning, cognitive flexibility, or functional fixation studies important aspects of creativity without using the word creativity.

Facilitation of creativity needs to activate both competencies, skills, and enabling conditions for creativity, as well as reconsider the effects of the creative activity, its reinforcing function, cognitively as well as socially, also in relation to what creativity accomplishes, the outcome of creativity [8]. Thus, the development of theory related reflexivity on creative experiences and competence as well as training of required skills in innovation processes needs to activate both the pre-requisites and predictors of creativity.

Three factors of creativity constitute criteria for assessment of creative responses in three dimensions 1) fluency, that is, number of responses/suggestions/ideas during a specified time period), 2) flexibility, that is, the number of different responses/suggestions/ideas during a specified time period and 3) originality, for example [10–12], is assessed on criteria of a qualitative nature based on perceptions of deviations from the ordinary [13, 14].

In general, people are expected to be flexible when encountering new situations, new requirements, and new problems, adapting to new technologies and communicating in diverse cultural settings. Cognitive flexibility is important for living, working, and learning in our rapidly changing world [15]. Social flexibility can be practiced and trained by giving and receiving information that develops new insights into other people's opinions and perceptions, it is the ability to be critically susceptible in social interaction and exchange for alternative possibilities [16, 17]. Social flexibility is thus required for effective teamwork and a facilitator of interpersonal communication [18]. A social perspective on flexibility can be defined as the inclination to adjust one's view to suit changing interpersonal situations, an interactional trait conceptualized as "openness to others." When interactors display flexible behavior, it is perceived as social flexibility [19].

Creativity involves the interplay of several factors where the correlation between social interaction processes and the characteristics of interacting individuals needs to be addressed together [20, 21]. The generation of original initiatives is a result of divergent thinking processes whereas the blending of several deviant initiatives and the assessment of this combinational outcomes' appropriateness is regarded as a convergent thinking process [22, 23].

Chrysikou [24] maintains that generative processes that evoke originality are characterized by spontaneous, emergent bottom-up processes, whereas convergent processes are controlled, top-down processes focusing on a particular goal or result, rather than producing original content.

For research on social interaction creativity, it is particularly interesting that divergent thinking and broadened attention are mutually interconnected with prosocial behavior [25, 26]. In addition, this broader attention divergence is mutually related to underlying generic cognitive processes for social interaction and decision-making, c.f., executive functions [27, 28].

2.1 Social interaction and exchange: Group creativity

The need for skills for social interaction and exchange does not diminish in a distributed way of working such as digital meeting tools. The digital meeting tools do not function as designers expect as long as the interactants (users, participants) do not understand what social interaction quality consists of and how dialog and exchange of ideas can be developed. Furthermore, in an increasingly digitized and automated working life, the abilities associated with creativity will be increasingly in demand [29, 30].

We first need a distinction regarding 'social group', a term which in the innovation literature is used interchangeably as 'organizational teams', and then define social interaction in that context. Definition of group: important distinction between dyadic versus group interaction, a group consists of more than two interactants [6], that is, we are not dealing with dyadic interaction. Definition of interaction: more than "performing some actions in synch with each other", that is, the quality that emerges during iterative exchange (such as in idea generation and creative problem solving) is related to the shared content and intersubjective understanding of that specific content. Definition of "interaction quality" [31, 32]:

There are different types of tasks and there are different levels of interaction quality as well. According to Sawyer [33] a work task complexity is defined in terms of how many operations that team members are required to work on together face-to-face, while a non-complex task can be performed more linearly without interacting face-to-face, such as via mail contact where each performs his part and sends to the next team member to add their part in the whole. An additive interaction characteristic is when every single group member's contribution is collected and put together to find out similar and deviant opinions to reach consensus. That is something that all members can agree upon—which often tends to be the lowest common denominator. But, to develop content through the pool of interaction and results or solutions based on all members' continuous contributions is not an additive process. And, from a group-creativity point of view the evolving content, the result, will probably be both new, unexpected, and useful as well as created from a genuine combination. Thus, it thereby meets all aspects of general definitions of creativity. In addition, Austin & Devin [34] used the ensemble concept to describe a specific quality of group interaction, often manifested in artistic ensemble interplay.

2.1.1 Group creativity

Group creativity can be described in terms of alterations between the individual's creativity (divergence) and developed shared group ideas (convergence). Group creativity includes both the production of new ideas and the stability to be able to integrate them into a solution or an innovation. The integration process of group ideas can be described in terms of circularity and emergence, at the collective level, ideas, and structures emerge through interactions between individuals and these collective structures influence the interactions between individuals. This iterative self-organizing group process is collective creative agency: an engagement of actors who, through the ongoing meaning-making of group members' initiatives and interplay of habit, imagination, and judgment, develops creative actions and decisions as an interactive response to the problem posed by change [4, 35].

This view on group interaction and idea exchange was developed in a study on working groups at five companies in mid-Sweden [36, 37]. To describe and present research results regarding group interaction and emerging group ideas, a diagrammatic visualization is well suited [38]. In these diagrams, the intensity of a group's exchange of ideas can be described and related to its capacity for self-organization and decision-making. In studies of group creativity, intensity can correspond to the concept of fluency at the individual level and the number of different group ideas during the same group session corresponds to the concept of flexibility at the individual level [36, 39].

2.2 Predictors of creativity or collaboration

2.2.1 Body movement

My research has involved artistic processes in both music and theater. The theater's methodology for establishing an ensemble of actors through physical movement and dance can be used as a tool to understand what creative processes in groups can be [40]. Artistic processes can also exemplify the potential of human interaction in terms of interactional qualities and abilities to listen and respond. One such example is how the collective ability to improvisation can support the self-organizing capabilities of groups [4, 41]. Another example of how rhythmic movements to music enhance school children's idea generation and problem-solving capabilities [42].

2.2.2 Originality takes abrasion

Creativity in work is the process of engaging practices that generate useful novelty. Organizations need to promote creative attitudes and provide open structures that can absorb resulting innovation. Core competencies in creative processes, such as critical thinking, problem-solving, and exchange, during teamwork are central to establishing distinctions between human-technology-artificial intelligence (e.g. [29, 30, 43]). Abilities that combine originality and usefulness in different ways characterize a creative agent. Original (different, deviant, unknown) actions need to be able to be used (meaningful, appropriate) functionally. Understanding and embracing this originality as appropriate and potentially valuable change is correspondingly an expression of creative agency capability. This creation and incorporation of new ways of doing things are not frictionless, the new is often perceived as questioning and criticizing, sometimes destructive. But the constructive contribution of creativity is to come up with alternatives to what it criticizes, this is the convergence of creative processes. Interactants, i.e., group members, need creativity for divergent differences to interact.

Creativity's laborious transcendence processes, balancing divergence and convergence, are often driven by enthusiasm and disruption—enthusiasm opens up to the new and discovering, while what threatens existing understandings and accustomed practices is perceived as disruptive. The new ideas criticizing the prevailing, that is, idea generation by definition means that all proposals and ideas cannot be combined. This critical function and even destruction as a result of the old needing to give way to the new is often overlooked aspect of creativity. Innovation seeks energy, drive, and an attractive work environment in creativity but is often surprised by criticism, slowness (incubation), and destruction. This ambiguous effect of creativity at the workplace utilizes idea generation to develop the workplace climate [44].

2.2.3 Dialog and intersubjective ideas

How creativity emerges over time in different types of workplace-specific situations is a central question. A result of creativity in social interaction is the development of shared understanding in relation to innovation capabilities through dialog seminar methodology [41]. Findings from this study contributed to the understanding of attitudes and approaches that establish and maintain group dialog that develops intersubjective understanding. A consequence was that intersubjective understanding, emerging group ideas need to be made explicit and visualized, primarily for the group members'/participants' sensemaking and re-understanding of what is going on and thus act constructively and contribute to the continued development of the group ideas [45].

With this perspective and purpose as a starting point, a conceptual framework was developed that later came to be known as "GroPro" including a number of associated "tools" for training the creative abilities of groups [4]. This project was driven by the belief that creativity and group creativity can be learned and trained [39].

2.2.4 Interaction quality: Group flow

The psychological experience called flow can occur during the performance of challenging activities in which the difficulty of the task is matched to the skill level of the person [46]. Characteristics of the flow experience include high but subjectively effortless attention, a sense of control, loss of self-awareness, and altered experience of time and enjoyment [47]. Flow experiences have predominantly been investigated in individual performers, there is a growing research interest in the quality of shared flow experience in social contexts, that is, group flow [48]. Sawyer [33] defines group flow as *"an optimal collective experience that occurs when members develop a feeling of mutual trust and empathy, in which individual intentions harmonize with those of the group."* When team members experience this synchronized state of flow, this can be considered a sign of increased performance and enhanced team-level effectiveness [49]. This type of self-reinforcing circularity between the group and its individuals is common in accounts of social interaction and group dynamics [2, 3, 50].

Van den Hout and colleagues [49] posit seven prerequisites and four predictors of team flow experiences. The prerequisites for team members' experiences of this collective quality of interaction are shared values and mutual recognition, (1) collective ambition and motivation. Alignment between members, (2) individual goals, and (3) the teams' goal. Team members' skill levels should be comparable, some of which are unique to each team member and (4) the team should be able to integrate those skills. Establishing communication and feedback (5) each team member develops broadened perspectives which thereby set the conditions for listening and exchanging, to agree on activities for achieving the common goal. Creating a safe environment requires the elimination of unnecessary and unacceptable risks while allowing for and acknowledging the possibility that any team member may fail, which in turn gives teams (6) the freedom they need to take necessary risks by making them feel it is safe to take action. Team members keep one another on task by using task-oriented behavior, accountability, and (7) mutual commitment to achieving the common goal [6, 16, 49, 51].

The four characteristics of team flow are (1) intense collaboration between team members as they strive towards the collective ambition and the achievement of their goals. Interpersonal concentration on the shared activity and holistic focus, a shared sense of unity, that the team has merged to form (2) a cohesion of blending egos. Team members' awareness of (3) constantly and effortless synergistic accomplishments, the willingness to be vulnerable and interdependent on one another, and finally shared confidence and (4) mutual trust that the team can achieve its common task [33, 49].

As Pels et al. [48] point out group performance should be assessed considering the specific task framing the situation and constitute the basic criteria relevant to the specific interaction quality. However, the differing collective aspects should be seen as separate pieces of a single puzzle that come together to form the whole. With this approach, the definition of group flow becomes the integration of several interacting factors, as follows: Group flow is a shared experience of states represented by (a) positive interactions, (b) a high collective competence, and (c) a collective state of mind consisting of positive relationships between group members, often resulting in (d) optimal collective performance and creativity [48].

2.2.5 Interaction quality: Flow synchronization

Flow synchronization is a psychological mechanism stimulating the group members to interact with each other, and to work on shared goals collaboratively to reach a challenging interdependent task [52]. This specific quality of interaction can develop when the interactants have experience working together in an intensive exchange of initiatives, ideas, and opinions [32].

Predictors of group members' experience of flow synchronization are (a) that they know the purpose of the task and (b) share a common strategy to reach agreed goals. Group members help each other, (c) integrate their initiatives with consistency, (d) motivate themselves, and (e) learn from each other. In addition, when they (f) reflect on the experience of working together, they realize how much they have developed during the activity and how they influenced each other's performance [53, 54]. Thus, the coordination effect of interactional functioning has been highlighted in the studies of flow in a social context.

The flow synchronization has been operationalized with the 28-item Flow.

Synchronization Questionnaire [52], which collects the experienced components of the interaction during a shared flow situation. The questionnaire identifies five components: (1) effective cooperation and partnership evaluate the common activity from a personal, relational perspective. The component of (2) engagement and concentration on the task refer to the flow experience during the interdependent activity. The third and fourth factor focus on the motivational effect of the partners, related to the concept of emergent motivation [46] and the facilitating role of the partners (e.g. [55]). Coordination during the interdependent activity refers to the behavioral coordination of the cooperative partners, supporting the synchronization mechanism [54].

2.3 Prerequisites for interaction quality

2.3.1 Openness

The personality trait Openness is one of five factors in the well-established framework called the Big-Five, which is used to assess individuals' personality characteristics. Each factor, i.e., trait, is based on empirically derived personality traits which in turn are a cluster of several more specific aspects which in turn comprise a large number of even more specific characteristics [15, 56].

The experience of being engaged in an activity depends mainly on the existing degree of openness. For example, group members who have an open attitude tend to experience a greater effect from several days of training activities in improvisational approaches [57]. In social groups, engaging in creative problem-solving tasks [58] and openness to experience includes intellectual curiosity, reasoning and imagination, artistic and esthetic motivation, as well as emotional and fantasy richness [15].

Openness to others is associated with creativity and flexibility in thinking and acting [19]. Flexibility in thinking is related to divergent thinking, such as "jumping" between categories to avoid functional fixation, i.e., getting stuck in predefined categorizations or "downpipe"-thinking. Openness is thus a prerequisite attitude for divergent and flexible as well as for improvisational, self-organizational, activities in a group [58].

2.3.2 Mindfulness/mindlessness

Personal characteristics such as the personality trait openness to experiences correlate with mindfulness, engagement, and concentration in collaboration and self-efficacy. Actively engaging in reconstructing one's impressions of the environment, for example, the behavior or suggestions of other group members and paying attention to what this may mean for one's understanding of what the group as a whole strives to create, for example, a group idea, is a mindful approach which is one of the prerequisites for collaborative creativity. This approach enables own interpretations and conscious manipulation which in turn can constitute creative contributions to the group's interdependent task ([59], p. 4). From a creativity perspective, Mindfulness is a composite of four components: (1) novelty seeking, (2) commitment, (3) producing novelty, and (4) flexibility.

Mindlessness on the other hand refers to when the individual mindlessly forms a cognitive commitment to the information and freezes its potential meaning. Alternative meanings or uses of the information become unavailable for active cognitive use [60].

Research on mindfulness in organizational contexts refers to this decidedly Western notion of mindfulness in terms of a socio-cognitive approach [61]. Weick and Sutcliffe are drawing extensively on Ellen Langer's research and describe mindfulness as a rich awareness of discriminatory detail generated by organizational processes [62]. Valdesolo et al. stress the benefit of training group members' perceptual sensitivity towards the other group members' actions as this promotes performance in interdependent tasks [60].

In this socio-cognitive perspective, mindfulness becomes central to the mutual creation of meaning [63]. A mindful perceptual sensitivity towards others can thus be regarded as a prerequisite for interaction and exchange in social groups. Langer uses the concept of "sideway perception" relating to interaction theory regarding considering interactants (group members) actions, that these actions are not arbitrary but always express meaning, cf. intersubjectivity, [64].

In addition, from a sociocultural perspective on creativity, an attitude of mindfulness group members can interpret interactants' actions as multifaceted with several different meanings [65, 66]. Prerequisites for collaborative creativity are then openness for others, flexibility for divergence exchange, mindfulness in interaction, intersubjective creation of meaningful creativity, and improvisational attitude.

2.3.3 Improvisational attitude

For making the group collaboration to eventuate into creativity all group members have to be present in whatever is going on, aiming their focus on whatever emerges out of the social interaction. This corresponds to "improvisational attitude" defined as "being mindfully in the now" [59]. When the interactants are aware that their collaboration continuously produces some emerging content, they can relate the group interaction to what the group produces as an outcome related to the task. We denote the emergent content "group idea" [67]. This implies that constructive and creative group effort is built on group members' awareness and ability to focus on whatever emerges through their interaction [68].

Sawyer outlines four rules or principles for constructive group improvisation [33]. Here follows our translation into four competence areas for collaborative creativity as a development of Sawyer's principles of group improvisation:

The basic competence for collaborative creativity is that the interactants train themselves to execute a "Yes, and.." attitude. The "Yes..."-part is the foundation of collective creative performance. The "and..."-part is the contributing dimension that supports the so-called "Group Idea" to gradually develop, that is, the emergent content.

The second competence area concerns listening skills. "Listen to the group idea!" is about listening empathetically to whatever is expressed rather than pondering about which response would be the most intelligent. This conscious awareness of others is consistent with one of Alex Osborn's guidelines for constructive brainstorming: "improve through combination" [69].

The third competence area is rather about developing a basic approach for all types of improvisation, that is, openness to others. A traditional concept in creativity research is functional fixation. The facilitative prompt for collaborative creativity aims to minimize this mindless fixation and reads "Do not write the script in your head!." When this call is heeded, interactants begin to be more mindful and "stay in the moment of interaction." It involves a great deal of trust in handing over control to the group process, assured that creativity will emerge from the iterative interaction. After all, it is not possible for one person to create a group idea.

Finally, the fourth competence area is about avoiding interrupting the group members' (interactants') synchronized experience of flow. That is, do not slow down the dialogic exchange of ideas and suggestions with long justifications and explanations, but "describe by doing" instead. The intensity, that is, fluency, of collaborative creativity has a greater impact on both performance and results than laborious accounts of the excellence of a particular idea.

2.3.4 The ability to combine differences and reaching alternatives

The responsibility for achieving and maintaining the presented prerequisites can to some extent be attributed to the group's leadership and distribution of work tasks, but the working group as a collective should also take responsibility for its way of interacting and making decisions, that is, the development of interaction maturity [31]. For the analysis of the interaction maturity of work groups, we have established a conceptual framework in the form of a maturity ladder inspired by Dreyfuss & Dreyfuss' [70] five levels of competence, as follows:

First level—the Novice: Group members' actions have no particular meaning for the other members' understanding of the task or problem being solved. Second level- the Beginner: Group members' action is collected to build consensus or to find the lowest shared denominator. Third level—the Experienced: The awareness of the group idea function as an interpretation background for members' action. Different actions could be understood and given a shared meaning about the group idea. Fourth level—the Competent: A shift in interaction quality from the lower levels. The way of relating to (attitude) group members' actions have an explicit connection to the evolving group idea and this understanding is the background from which members' action is interpreted and given meaning. Fifth level—the Expert: The team is able to completely change the way of interaction. The team realizes and refers explicitly to the inter-subjective group idea and is able to shift between different group ideas [31].

3. Research approaches, conceptualization, and design

For new research initiatives to contribute to the development and expansion of the research field of collaborative creativity, it is important to relate to a common conceptual framework. Research on creativity within psychology-oriented scientific perspectives and methodologies such as psychodynamic, cognitive, and behavioral sciences. Research on creativity with cognition perspectives has developed methodology and conceptual frameworks for studying creativity (e.g. divergent thinking, functional fixation, incubation, traits), where, for example, divergent thinking is operationalized and analyzed in three conceptual dimensions of fluency, flexibility, and originality.

Research with social psychology-oriented perspectives has developed methodology and other types of conceptual frameworks for studying creativity (e.g. social interaction, interdependence, group dynamics). Research findings on socially oriented creativity should also be related to divergence and convergence as well as originality and functionality.

To establish a common framework to enable comparison of findings, we suggest the following conceptual analogies: *Fluency* in studies on creativity in group interaction should be assessed and analyzed in terms of "intensity" in the exchange of ideas and dialog. *Flexibility* in group interaction should be assessed and analyzed in terms of "the number of different types of group ideas" that the group is able to accommodate during group interaction. The creativity dimension *originality* should be assessed and analyzed in relation to the conventions in the relevant field. For example, an originality in group interaction can be assessed and analyzed in terms of deviations from expected behavior, such as the degree of collective improvisation which the group cohesively manages to act and treat.

3.1 Research conceptualization and design

To form a coherent and consistent development through the sections of this chapter from the theory-based conceptual framework, via research approaches and design, a model for research design on collaborative creativity is presented in this section. This has the overall purpose of enabling comparable studies in terms of methodology and findings to continue developing the knowledge area of collaborative creativity with joint efforts.

A scientific methodological approach that lends itself well to research on collaborative creativity is critical realism (4). Utilizing ontologically based stratification of structures (e.g. outcome, group ideas), events (e.g, idea generation, dialog), and underlying mechanisms (e.g. interaction quality, improvisational attitude), the combination of different types of methods for data collection and analysis is enabled, that is, retroduction [71]. In the research design according to **Figure 2**, stratification is visualized in four levels, where each level can consist of several events and each event can be caused by several mechanisms. It is the research question that determines which mechanisms a particular study should consider as appropriate explaining factors for the particular event which are thus analyzed through critical realism's so-called retroduction.

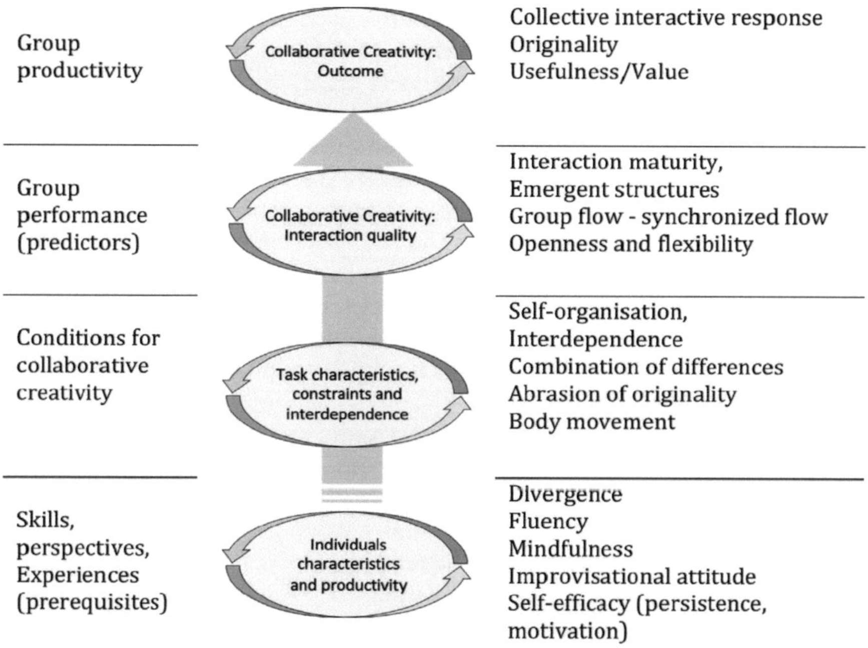

Figure 2.
Suggested research design for collaborative creativity; stratification levels, theoretical concepts for qualitative and statistical analysis.

In relation to the methodology and research approach of critical realism, the proposed research design, the *mechanisms* are grouped into predictors and prerequisites. The *events* have also been grouped into two categories, that is, "conditions for collaborative creativity" and "group productivity."

4. Discussion and conclusion

In the previous section, a methodological approach to investigating collaborative creativity was presented. The rationale for this is that creativity is composed of various factors that directly or indirectly influence each other, and these interdependencies are rarely simple and linear. Furthermore, the translation of traditional analysis concepts (i.e. fluency, flexibility, and originality) to social collaborative creativity, needs to be established to continue developing creativity research cumulative knowledge expansion by enabling comparison between the results of different studies.

It is, for example, important that findings in creativity research do not conflate creativity as outcome versus creativity as quality—or do not present a clear distinction between outcome versus performance. This can lead to problems in the assessment and analysis of collected empirical data and thus, by extension, the comparison of results between different studies, this especially applies to the originality dimension of creativity. Assessment of originality in creativity can be carried out by external experts grading the results of creative processes in the originality dimension continuum between everyday/traditional to unique/path-breaking, cf., consensual assessment technique, CAT [13]. However, assessment and analysis of the originality dimension in terms of creative quality in the process or more specifically in the group interaction is not as developed as the other dimensions, that is, fluency and flexibility.

In **Figure 2**, it was demonstrated that systematic research studies need to stratify these factors ontologically at different levels to understand, analyze and explain how factors at different levels interact in collaborative creativity and establish conditions that drive interaction quality and productivity.

Moreover, factors at different levels may need different methods of collecting and analyzing data to relate results between different levels, which is supported by a theory-based conceptual framework. Each with appropriate methods for data collection and analysis of mechanisms that at different levels contribute to the emergence of collaborative creativity. The construction of a conceptual framework for research should therefore consider and translate central concepts consistently between levels, enabling comparisons of results between different studies. The suggested conceptual framework for collaborative creativity thus has its function both within and between different research studies (**Figure 3**).

This chapter has presented a social perspective on collaborative creativity evolving through the combination of divergent differences. The premise for developing an understanding of the elements of collaborative creativity is the circular logic captured in two propositions, P1: Interaction drives creativity, and P2: Creativity enables interaction. These propositions express the relational logic that "the interactants need creativity for divergent differences to really inter-act." Proposition P1 has been shown in research studies to have certain predictors, while proposition P2 has been shown to need the support of certain prerequisites.

Predictors of creativity:

Social interaction always establishes some kind of emergent property, in collaborative creativity, a specific interaction quality conceptualized as flow synchronization has been shown to be an emergent factor.

Idea generation develops Divergent thinking, which, in turn, initiates Executive functions. An example is the relationship between idea generation and creative productivity, which develops openness also in social interaction, such as empathic understanding of others' perspectives and perceptions.

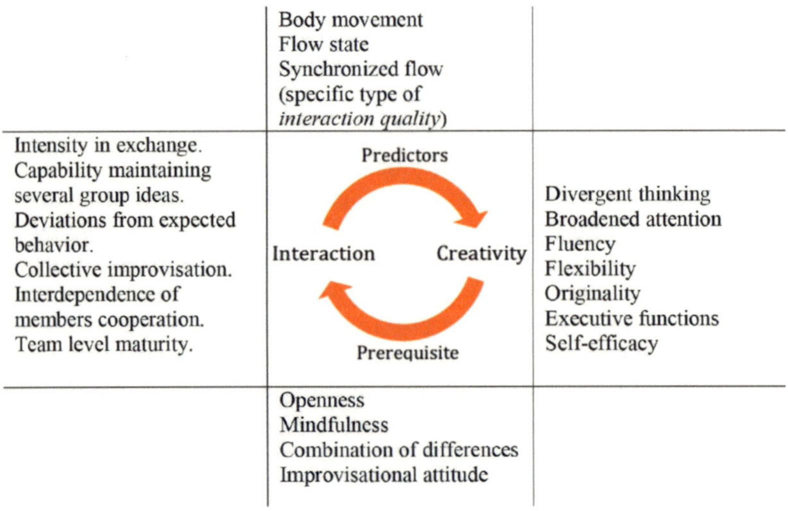

Figure 3.
Collaborative creativity—predictors of creativity-related prerequisites for interaction.

It is crucial that the Originality dimension is assessed especially in studies on creative group interaction because it is necessary to get results relevant to creativity. But it is also methodologically important to avoid the problematic concept of "dynamics", which seems to be the case when only the dimensions of Fluency and Flexibility are assessed and analyzed. It risks establishing yet another black box hiding what flexibility and fluency in social interaction can be also making comparison to other studies cumbersome, which in turn risks limiting the development of the research area that constitutes collaborative creativity.

Prerequisites for interaction:

An approach characterized by Mindfulness seems to work mutually reinforcing with Openness to others in a way that strengthens integrative action, that is, interaction between, for example, group members. These are two prerequisites for collaborative creativity. Two others are the desire to combine divergent differences and improvisational approach. These four have been shown to be prerequisites for collaborative creativity because they both support creative productivity and exchange in socially interdependent collaboration.

The presented premise should also constitute the pedagogical logic for didactic strategies that aim to activate predictors and prerequisites of collaborative creativity through the training of necessary skills and knowledge development. Acquiring these abilities and establishing relationships requires training and knowledge development. Education, training, and facilitation of collaborative creativity need to strive to structure learning and activate abilities based on the ontological prerequisites of creativity. The structuring of content should be connected in such a way that the students have the opportunity to be activated in creative productivity, analyzing the outcome of interaction and exchange, reflecting on flow experience and interaction quality in relation to the emergent content and evolving structures.

5. Conclusions

The presented reasoning and developed model of research design for investigation of collaborative creativity are derived from both qualitative and quantitative methodological approaches. The chapter advances the view that the impact of activities training group-based collaborative creativity should be elevated in importance beyond individuals' capabilities related to idea generation and production of original and functional ideas. Both training in and research on these abilities are important, but these activities should also be understood and developed based on their generative function in social contexts. Creativity is a genuinely generative ability, creative processes develop the ability to take initiative, make decisions, and interact constructively together and this should have implications for research approaches and methodological design as our conceptual framework for collaborative creativity suggests.

Practitioners, educators, and facilitators of creative processes should acquire a deeper understanding of the predictors and prerequisites of creativity. This recommendation is based on the presented research design, moreover, the conceptual framework are motivated by the fact that the social value of creativity is too often overlooked, that is, the combination and interaction between divergent differences, fostered understanding of different perspectives, diversity, and empathy. And, that this, therefore, needs to be taken more seriously into account in different types of education and training in creativity.

Educational initiatives for collaborative creativity should train skills that are needed as prerequisites for this type of interaction quality to be established. Such training should include assessment of creativity at different levels, that is, individual, group, workplace, etc. In addition, the development of facilitation skills should include the development of action plans for different types of method used based on the stratification of levels in the suggested research design (**Figure 2**) [72]. These learning activities are important for facilitators of collaborative creativity where expectations of both deep theoretical knowledge and experience-based abilities are addressed [7].

6. Future development

Based on the previous discussion, further development of collaborative creativity is proposed in the following three points:

- The development of research on creativity with sociological research approaches is promising, presenting studies often based on conceptual frameworks from action theory [64, 65]. From a sociological research perspective, collaborative creativity can be defined as a temporal interactive engagement between actors in different structural environments where the interaction both reproduces and transforms emergent content and responses such as responses to interdependent problem-solving tasks [35].

- In addition to mapping the methodologies of similar research approaches and their presented results with each other, the suggested conceptual framework needs to be applied in several studies to establish significance and create credibility.

- Research on collaborative creativity should focus on the effects of creativity rather than on the immediate results of creative processes since social interaction is needed for creativity as well as creativity is needed for social interaction, integrating differences, and intersubjective meaning-making.

Conflicts of interest

The author declared no potential conflicts of interest with respect to the research, authorship, and/or publication of this chapter.

Author details

Bengt Köping Olsson
Mälardalen University, Eskilstuna, Sweden

*Address all correspondence to: bengt.koping.olsson@mdu.se

References

[1] Tanggaard L. Creativity Must be Learned—When Talent Turns into Innovation. Aalborg: Aalborg University Press; 2008

[2] Ryan LV, Rutherford MA. Mary Parker Follett: Individualist or collectivist? Or both? Journal of Management History. 2000;**6**(5):207-223. DOI: 10.1108/13552520010348362

[3] Harwood J. Ludwik fleck and the sociology of knowledge. Social Studies of Science. 1986;**16**(1):173-187. DOI: 10.1177/030631286016001009

[4] Backström T, Söderberg T. Self-organisation and group creativity. Journal of creativity and business. Innovations. 2016;**2**:65-79

[5] Moreland RL, Levine JM. Group dynamics over time: Development and socialization in small groups. In: McGrath JE, editor. The Social Psychology of Time: New Perspectives. Newbury Park, CA: Sage; 1988. pp. 151-181

[6] Katzenbach JR, Smith DK. The Wisdom of Teams: Creating the High-Performance Organization. Boston: Harvard Business School; 1993

[7] Goller I, Bessant J. Creativity for Innovation Management. New York: Routledge; 2017

[8] Organization for Economic Cooperation and Development. Framework for the Assessment of Creative Thinking (Third Draft). Paris: OECD publishing; 2019

[9] Sternberg RJ, Lubart TI, Kaufman JC, Pretz JE. Creativity. In: Holyoak KJ, Morrison RG, editors. Cambridge Handbook of Thinking and Reasoning. Cambridge: Cambridge; 2005. pp. 351-370

[10] Torrance EP. Empirical validation of criterion-referenced indicators of creative ability through a longitudinal study. Creative Child and Adult Quarterly. 1981;**6**:136-140

[11] Beghetto RA, Kaufman JC. Do we all have multicreative potential? ZDM Mathematics Education. 2009;**41**:39-44

[12] Hocevar D. Ideational fluency as a confounding factor in the measurement of originality. Journal of Educational Psychology. 1979;**71**(2):191-196. DOI: 10.1037/0022-0663.71.2.191

[13] Amabile TM. Social psychology of creativity: A consensual assessment technique. Journal of Personality and Social Psychology. 1982;**43**(5):997-1013. DOI: 10.1037/0022-3514.43.5.997

[14] Sternberg RJ, Kaufman JC. The Nature of Human Creativity. Cambridge: Cambridge; 2018

[15] Kaufman SB. Opening up openness to experience: A four factor model and relations to creative achievement in the arts and sciences. Journal of Creative Behavior. 2013;**47**(4):233-255. DOI: 10.1002/jocb.33

[16] Hare AP. Creativity in Small Groups. Beverly Hills: Sage; 1982

[17] McCrae RR, Sutin AR. Openness to experience. In: Leary MR, Hoyle RH, editors. Handbook of Individual Differences in Social Behavior. New York: The Guilford Press; 2009. pp. 257-273

[18] McComb SA, Green SG, Compton DW. Team flexibility's relationship to staffing and performance in complex projects: An empirical analysis. Journal of Engineering and Technology Management. 2007;**24**(4):293313

[19] Barak M, Levenberg A. A model of flexible thinking in contemporary education. Thinking Skills and Creativity. 2016;**22**(2016):74-85. DOI: doi.org/10.1016/j.tsc.2016.09.003

[20] Rhodes M. An analysis of creativity. The Phi Delta Kappan. 1961;**42**(7):305-310

[21] Diakidoy I-AN, Constantinou CP. Creativity in physics: Response fluency and task specificity. Creativity Research Journal. 2001;**13**(3-4):401-410. DOI: 10.1207/S15326934CRJ1334_17

[22] Shamay-Tsoory SG, Adler N, Aharon J, Perruy D, Mayseless N. The origins of originality: The neural bases of creative thinking and originality. Neuropsychologia. 2011;**49**:178-185. DOI: 10.1016/j.neuropsychologia.2010.11.020

[23] Guilford JP. Creativity. American Psychologist. 1950;**5**(9):444-454

[24] Chrysikou EG. Creativity in and out of (cognitive) control. Current Opinion in Behavioral Sciences. 2019;**27**:94-99. DOI: 10.1016/j.cobeha.2018.09.014

[25] Wronska MK, Kolanczyk A, Nijstad BA. Engaging in creativity broadens attentional scope. Frontiers in Psychology. 2018;**9**:1772. DOI: 10.3389/fpsyg.2018.01772

[26] Colzato LS, van den Wildenberg WPM, Hommel B. Increasing self-other integration through divergent thinking. Psychonomic Bulletin Review. 2013;**20**:1011-1016. DOI: 10.3758/s13423-013-0413-4

[27] Mukherjee S, Srinivasan N, Kumar N, Manjalay JA. Perceptual broadening leads to more prosociality. Frontiers in Psychology. 2018;**9**:110

[28] Andersson P. Assessment and development of executive function (EF) during childhood. Child Neuropsychology. 2002;**8**(2):71-82. DOI: 10.1076/chin.8.2.71.8724

[29] Sowa K, Przegalinska., and Ciechanowski, L. Cobots in knowledge work: Human—AI collaboration in managerial professions. Journal of Business Research. 2022;**125**:135-142. DOI: 10.1016/j.jbusres.2020.11.038

[30] Kolade O, Owoseni A. Employment 5.0: The work of the future and the future of work. Technology in Society. 2022;**2022**:71. DOI: 10.1016/j.techsoc.2022.102086

[31] Köping Olsson B. Creativity and maturity in team-based innovation—A model for assessment of interaction quality in relation to task characteristics. In: The 1st ARTEM Organizational creativity International Conference. Nancy, France; 2015, 2015

[32] Köping Olsson B, Harmat L. The impact of creativity, flow, and interaction quality on collaborative design solutions in social groups. Journal of Creativity and Business Innovation. 2022;**8**(1):24-49

[33] Sawyer RK. Group Genius: The Power of Collaboration. New York: Basic Books; 2017

[34] Austin R, Devin L. Artful Making: What Managers Need to Know about How Artists Work. New Jersey: Financial Times Prentice Hall; 2003

[35] Emirbayer M, Mische A. What is agency? American Journal of Sociology. 1998;**103**(4):962-1023. DOI: 10.1086/231294

[36] Backström T, Moström Åberg M, Köping Olsson B, Wilhelmson L, Åteg M. Manager's task to support integrated autonomy at the workplace: Results from an intervention. *International*. Journal of Business and Management. 2013;**8**:22

[37] Wilhelmson L, Moström Åberg M, Backström T, Köping Olsson B. Enabling transformative learning in the workplace—An educative research intervention. Journal of Transformative Education. 2014;**13**(3):219-238

[38] Köping Olsson B, Florin U. Idea exchange and shared understanding: Tools stimulating thought and conveying ideas. Design Principles and Practices: An International Journal. 2011;**5**:487-502

[39] Köping Olsson B, Backström T. Innovative leadership—Supporting creative team interaction. In: Paper presented at The 7th International Symposium on Management of Technology. Hangzhou, P. R. China; 2012, 2012

[40] Bozic N, Köping Olsson B. Culture for radical innovation: What can business learn from creative processes of contemporary dancers? Organizational Aesthetics. 2013;**2**(1):59-83

[41] Backström T, Köping Olsson B. Kaikaku—A complement to emergence-based development. In: Paper presented at The First conference on Design Creativity. Kobe, Japan: Kobe International Conference Center; 2010

[42] Österberg P, Köping Olsson B. Dancing: A strategy to maintain schoolchildren's openness for idea generation. Journal of Physical Education, Recreation & Dance. 2021;**92**(3):20-25

[43] Rampersad G. Robot will take your job: Innovation for an era of artificial intelligence. Journal of Business Research. 2020;**116**:68-74. DOI: 10.1016/j.jbusres.2020.05.019

[44] Köping Olsson B, Carlsson A-L. Creativity practice, an angel or devil in the innovation eden: The function of criticism and destruction

[45] Andersson Schaeffer J, Köping Olsson B. Spaces for creative dialogue—Context and content in practice. In: Conference Contribution Innovating the Experience Economy, Design, Consumption and Concepts. June, Roskilde, Denmark; 2012. pp. 11-12

[46] Csikszentmihalyi M, Csikszentmihalyi IS, editors. Optimal Experience. Psychological Studies of Flow in Consciousness. Cambridge, UK: Cambridge University; 1988

[47] Csikszentmihalyi M, Nakamura J. Effortless attention in everyday life: A systematic phenomenology. In: Bruya B, editor. Effortless Attention: A New Perspective in the Cognitive Science of Attention and Action. Cambridge, MA: The MIT Press; 2010. pp. 179-190

[48] Pels F, Kleinert J, Menningen F. Group flow: A scoping review of definitions, theoretical approaches, measures, and findings. PLoS One. 2018;**13**(12):e0210117. DOI: 10.1371/journal.pone.0210117

[49] van den Hout JJJ, Gevers JMP, Davis OC, Weggeman MCDP. Developing and testing the team flow monitor (TFM). Cogent Psychology. 2019;**6**:1-27. DOI: 10.1080/23311908.2019.1643962

[50] Boje DM, Rosile GA. Where's the power in empowerment? Answers from Follet and Clegg. The Journal of Applied Behavioral Science. 2001;**37**(1):90-117. DOI: 10.1177/0021886301371006

[51] Gaggioli A, Mazzoni E, Milani L, Riva G. The creative link: Investigating the relationship beween social network indices, creative performance and flow in blended teams. Computers in Human Behavior. 2013;**42**:157-166. DOI: 10.1016/j.chb.2013.12.003

[52] Magyaródi T, Oláh A. Flow synchronization questionnaire: Measuring the mechanism of optimal experience in interactions. Mentálhigiéné és Pszichoszomatika. 2015;**16**(4):271-296

[53] Magyaródi T, Oláh A. The effect of social interaction on flow experience. International Journal of Psychology and Behavior Analysis. 2017;**3**:126

[54] Magyaródi T, Köping Olsson B, Harmat L. Flow synchronisation and shared emotions towards understanding collective flow experience. In: Arts and Mindfulness Education for Human Flourishing. London: Routledge; 2022

[55] Salanova M et al. Flowing together: A longitudinal study of collective efficacy and collective flow among workgroups. The Journal of Psychology. 2014;**148**(4):435-455. DOI: 10.1080/00223980.2013.806290

[56] Gosling SD, Rentfrow PJ, Swann WB Jr. A very brief measure of the big-five personality domains. Journal of Research in Personality. 2003;**37**:504-528. DOI: 10.1016/S0092-6566(03)00046-1

[57] Kaufman SB, Quilty LC, Grazioplene RG, Hirsh JB, Gray JR, Peterson JB, et al. Openness to experience and intellect differentially predict creative achievement in the arts and sciences. Journal of Personality. 2016;**84**(2):248-258. DOI: 10.1111/jopy.12156

[58] Köping Olsson B. Covarying effects of creativity studies on academic performance in innovation processes. In: Proceedings of the XXXIII ISPIM Innovation Conference 'Innovating in a Digital World'. Copenhagen, Denmark; 2022

[59] Langer EJ. Mindfulness. Cambridge, MA: De Capo Press; 1989

[60] Valdesolo P, Ouyang J, DeSteno D. The rhythm of joint action: Synchrony promotes cooperative ability. Journal of Experimental Psychology. 2010;**46**:693-695. DOI: 10.1016/j.jesp.2010.03.004

[61] Pirson MA, Langer E, Zilcha S. Enabling a Socio-cognitive Perspective of Mindfulness: The Development and Validation of the Langer Mindfulness Scale. Journal of Adult Development. 2018;**25**:168-185. DOI: 10.1007/s10804-018-9282-4

[62] Weick KE, Sutcliffe KM. Mindfulness and the quality of organizational attention. Organization Science. 2006;**17**:514-524. DOI: 10.1287/orsc.1060.0196

[63] Langer EJ. On Becoming an Artist: Reinventing yourself through Mindful Creativity. New York: Ballantine Books; 2006

[64] Pina-Cabral J. Modes of participation. Anthropological Theory. 2018;**18**(4):435-455

[65] Glaveanu VP, Gillespie A, Valsiner J, editors. Rethinking Creativity: Contributions from Social and Cultural Psychology. 1st ed. New York: Routledge; 2015

[66] Glăveanu VP, Gillespie A, Karwowski M. Are people working

together inclined towards practicality? A process analysis of creative ideation in individuals and dyads. Psychology of Aesthetics, Creativity, and the Arts. 2019;**13**(4):388-401. DOI: 10.1037/aca0000171

[67] Köping Olsson B. Descriptive languages in and for creative practice: Idea development during group sessions. In: Swedish: Beskrivningsspråk i och för kreativ praxis: Idéutveckling under gruppsession. Västerås, Sweden: Mälardalen University Press; 2008

[68] Sawyer RK. Creating Conversations: Improvisation in Everyday Discourse. New Jersey: Hampton Press; 2001

[69] Stein MI. Stimulating creativity, Volume 2: Group Procedures. New York: Academic Press; 1975

[70] Dreyfus HL, Dreyfus SE. Mind over Machine. The Power of Human Intuition and Expertise in the Era of the Computer. London: Blackwell; 1986

[71] Danermark B, Ekström M, Karlsson JC. Explaining Society: Critical Realism in the Social Sciences. Abingdon, Oxon: Routledge; 2019

[72] Plucker J, Beghetto R, Dow G. Why isn't creativity more important to educational psychologists? Potentials, pitfalls, and future directions in creativity research. Educational Psychologist. 2004;**39**(2):83-96

Chapter 3

Bisociation: Creativity of an Aha! Moment

Bronislaw Czarnocha and William Baker

Abstract

This chapter presents a new theory of creativity in mathematics education. The theory has been anchored in two foundations: that of teaching practice of Vrunda Prabhu which occasioned surprisingly many Aha! Moments experienced by students in her remedial arithmetic classes in college and that of the Arthur Koestler's volume The Act of Creation. The Act of Creation introduces term bisociation describing an Aha! Moment and Eureka Experience as "the spontaneous leap of insight which connects previously unconnected frames of reference" by unearthing "hidden analogies". Whereas Koestler formulated the concept of bisociation within humor, scientific discover and art, we focus primarily on the bisociative creativity in mathematics. We abstract the concepts of the bisociative frame as the two unconnected frames of reference, useful method for (1) identification of heightened probability of creative insight in a given mathematical situation and (2) identification of possible creativity within different theories of learning. The chapter explores Koestler/Prabhu theory of learning through Aha! Moment and applies the bisociative frame to investigate its processes of interaction with several different theories of learning. Uncovered processes of interaction suggest that creativity should be the basis of contemporary learning and teaching.

Keywords: creativity, Aha! Moment, bisociative frame, interiorization, internalization, appropriation, Koestler/Prabhu theory of learning, constructivist, socio-cultural

1. Introduction

This chapter presents a new theory of creativity in mathematics education. The theory has been built from two sources: Prabhu's [1] teaching experiment, which brought about a surprising number of Aha! Moments in students enrolled in her developmental math classes and Arthur Koestler's volume The Act of Creation [2]. The research, or rather the teaching–research pathway that led us to formulating the new theory of creativity in mathematics education, started during the 2010/2011 teaching-experiment conducted by Prabhu and supported by CUNY grant C^3IRG 7, *Problem Solving in Remedial Mathematics—a Jumpstart for the Reform*. The grant was awarded to members of the Teaching-Research (TR) Team of the Bronx anchored in Hostos Community College, New York City. The work on creativity of Aha! Moment took two creative cycles of work each lasting several years, during which we realized we have here a new distinct approach to creativity through Aha! Moment. Since Aha! or epiphanies are common occurrences among the general public in many domains of activity, the presented theory applies equally to the "underserved and underrepresented" as well as to the "talented and gifted".

These separate research ideas coalesce into a whole presented in the work by Czarnocha, Baker, Dias and Prabhu in [3], where the first sketches of the new theory appeared in different chapters. The volume introduced the theory of creativity in its discussion of teaching–research methodology, showing that TR methodology is intrinsically creative. Because teaching and research constitute two matrices of thought, generally and unfortunately not connected with each other, working at the teaching–research interface enables bisociative processes to create/build Aha! bridges between them.

During the second teaching–research creative cycle, we investigated processes of facilitation of an Aha! Moment assessment of the depth of knowledge gained during the insight, and constraints imposed on development of creativity in mathematics classrooms. The full exploration of the phenomenon Aha! Moment took place at the completion of the second creative teaching–research cycle with the volume Creativity of an Aha! Moment and Mathematics Education, Czarnocha [4] which collected our own research as well as that of international experts on Aha! Moments. The Koestler/Prabhu theory of creativity through Aha! Moment was formulated in that volume for the first time. The reflection on this work provided the basis on which to build the Philosophy of Creativity in Mathematics Education [5]. In this chapter, we present the Koestler/Prabhu theory in Section 2 together with some of its application into creativity research. We also provide a method of assessment for the increase of knowledge occasioned by Aha! Moment insight in Section 3. Section 4 presents our investigations into the relationship between the bisociative frame and sociocultural framework. In Section 5, we summarize and connect different threads of the paper.

2. Koestler theory of bisociation

2.1 Elements of the Koestler theory of bisociation

The work presented here on the creative moments of insight popularly called Eureka experience or Aha! Moment is based on [2] where the author introduced a new concept/term of bisociation, which is the spontaneous act of thought which combines or dialectally synthesizes the information from two different, generally unconnected domains. Bisociation is seen here as distinct from association, which produces knowledge within a single domain. Koestler makes a clear distinction between more routine or habitual thinking (association) operating within a single plane or a matrix of thought, referred to here as the exercise of understanding and the more creative bisociative mode of thinking that connects independent autonomous matrices called progress in understanding [6].

What is a matrix of thought also called a frame of reference?

> *The matrix is the pattern before you, representing the ensemble of permissible moves. The code which governs the matrix can be put into simple mathematical equations … or it can be expressed in words. The code is the fixed invariable factor in a skill or*

> *habit, the matrix is the variable part. The two words do not refer to different entities, they refer to different aspects of the same activity.* ([2], p. 40).

A clear example of a matrix and a code is given in chess where the matrix is the full collection of available moves, while the code are the rules for the movement and interaction among the chess pieces. In high school algebra, the matrix is the full set of polynomials and power series in one or two variables; the code are the established rules of operations on these mathematics objects.

In our effort to apply Koestler's theoretical framework in the creativity process within learning mathematics, we focus on two related questions: (1) How can you describe the genesis of a new code-matrix? and (2) How can you characterize moments of creative insight that lead to new structure-matrices and ultimately new codes? Koestler considered the formation of new structure-matrices, as well as the hierarchy of matrices serving the organism as "[s]ymbolic models of the external world" (p. 506) that govern how we interpret and react to a situation in a predictable manner, as the central role of cognition. For Koestler, an important component in characterizing moments of insight is expressed in his notion of the degree of originality, which is inherent in our discussion of the depth of knowledge acquired during a moment of insight. In learning theory, an individual's development of structure is a focal point of constructivist research and cognitive science.

Koestler translates the term matrix very broadly into such diverse domains as literature, art, drama, motor skills and humor. He notes that matrix in cognitive psychology would be termed a schema.

> *The concept of matrices with fixed codes and adaptable strategies, proposed as a unifying formula, appears to be equally applicable to, perceptual, cognitive, and motor skills and to the psychological structures variously called, 'frames of reference', 'associate contexts', 'universes of discourse', 'mental sets', or 'schemata' etc.* (p. 96)

Cognitive psychologists use the term schema to describe mental structures that guide our response to life situations; thus, there is a schema for work, family relationships, commuting, religious services, etc. The particular schema that interests us is an individual's schema for mathematics problem-solving.

Constructivists use the term scheme to describe a mental process for resolving problem situations. We loosely translate schema as a hierarchical structure of schemes. Thus, the building of schema begins in the second stage of the Piaget Garcia Triad [7] through the connection of schemes into hierarchical collections (mental toolboxes for a given domains of math). The third stage begins connections between schemata. The term code translates into the constructivist notion of an invariant relationships, which can be understood as the automated principle of a conceptual relationship that underlies an activity–effect relationship in a problem situation (scheme).

Creativity or a Eureka moment occurs when an idea is suddenly understood to exist simultaneously in two previously unrelated frame (**Figure 1**).

> *The perceiving of a situation, or idea, L, in two self-consistent but previously incompatible frames of reference (fig. 1) The event L in which the two intersect is made to vibrate simultaneously on two different wavelengths, as it were. While this unusual situation lasts, L, is not merely linked to one associative context, but bisociated with two.* ([2], p. 35).

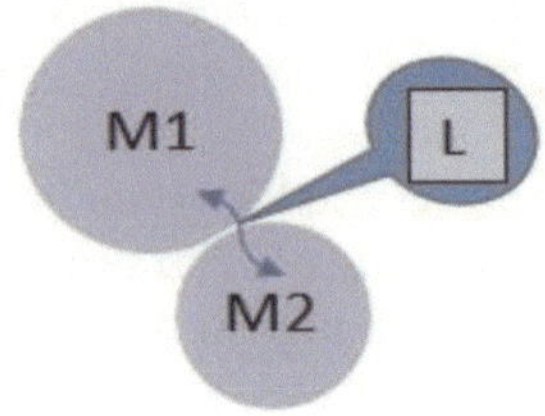

Figure 1.
Bisociated concept L between matrices M1 and M2.

2.2 Koestler: progress in understanding new codes

In the first stage of the Piaget Garcia Triad, the birth of a conceptual relationship begins with interiorization. For Koestler the genesis of a new code is referred to as progress in understanding; the equivalent constructivist term is accommodation. In constructivist theory, all acts of accommodation are the result of what Piaget refers to as reflective abstraction (abstraction of processes). Interiorization is understood as the foundational type or process of such abstraction that translates externally directed activity into an internal process. Koester describes the genesis of codes, or progress in understanding, as due to moments of insight, within the learning process, or moments of creativity insight that lead to the synthesis of two matrices or, "bisociation"

> *[p]rogress in understanding – the acquisition of new insights ... is achieved by the formulation of new codes by ... empirical induction, abstraction, and discrimination, bisociation*. (p. 619).

The Transition from Empirical to Abstract reasoning.

Concepts are born in the first stage of the Piaget & Garcia Triad. Constructivist research based on Piaget's work and social constructivism based on Vygotsky; both view the development of a child's cognition as transitioning from empirical to more abstract reasoning. Constructivists view the transition from empirical reasoning, during which one's intuitive solution activity is directed by the situation, as interiorization. This process leads to an abstract activity–effect relationship based on understanding the relevant concept and thus no longer dependent on the situation. They view this first-stage transition as the way human cognition evolved historically as well as the pathway for child development.

Vygotsky refers to the transition from empirical to abstract thought as one from spontaneous to scientific concepts and considers it essential for the dawning of the child's ability to engage in self-reflection. Vygotsky considers such internalization of exterior activity as social-based, primarily guided by adult communication, and in large part based on imitation of adult behavior.

In this debate, Koestler is strictly a constructivist. Indeed in his chapter on science and emotion, he titled one of the sections "The Boredom of Science" lamenting how direct instruction of theorems and repetition have made math and science "antihuman". In contrast to the social-cultural approach, in his view individual moments of insight lead to a transcendence experience, and only from this can we begin to appreciate math and science. Like Vygotsky, Koestler understands objective abstract thought as essential for human self-reflection, but like constructivists he situates such reflection, as based upon transcend moments of insight that are individualistic. Thus,

Koestler like constructivist researchers is primarily focused on bisociation within the learning process of an individual and not on moments of insight within social discourse.

The phenomenon under investigation is moments of insight within the learning process, which lead to progress in understanding, or concept development. As teacher researchers our contention is that such moments occur both during social discourse-internalization as well as during reflection on our own solution activity. Furthermore, the novel concepts and structure developed through the connections established at this moment of insight (progress in understanding) can be assessed through the Piaget-Garcia Triad. Before continuing with these themes, we review Eureka or Aha! Moments and creativity theory.

2.3 Eureka experience/Aha! Moment

The phenomenon under investigation of the proposed theory is the act of creation in mathematics and science called Eureka experience or Aha! Moment. It is that moment when suddenly, after a long period of trying to solve a problem or understand a new concept without success, the solution comes in a flash, generally with a good doze of satisfaction. It is a very particular form of creativity that appears as an insight, as a discrete insight in that it appears instantaneously at separate moments of time.

Gestalt creativity approaches the Eureka experience as the stage of illumination within the sequence of preparation, incubation, illumination and verification stages suggested by [8, 9]. The sequence represents the stages through which the formation of the creative idea takes place. While the first three stages came from psychological research, the fourth stage was added as the necessary part of the creative process at the Poincare insistence [8]. Why would Poincare insist on verification as the components of creativity? Because Poincare in [10] says: *It never happens that unconscious work supplies a ready-made result of a lengthy calculation in which we have only to apply the fixed rules ... All that we can hope from these inspirations, which are the fruits of unconscious work, is to obtain points of departure for such calculations. As for calculations themselves, they must be made in the second period of conscious work, which follows the inspiration and in which the results of inspiration are verified and the consequences deduced* (pp. 62–63).

The more so, of course, because sometimes the Aha! Moment is false. Consequently, the verification stage plays a dual role: On one hand as a possible completion of the creative act and as the check on the correctness of the logical-causal structure of its content. Recent examination of Wallas's work suggests in [11] the fifth stage of intimation to be placed between incubation and illumination. Intimation is, for Wallas, the

> *"fringe of consciousness" which surrounds our "focal" consciousness as the Sun's 'corona' surrounds the disk of full luminosity", Wallas continues: "This fringe consciousness may last up to the flash instance, may accompany it, and in some cases may continue beyond that."*

In such a case intimation can be an excellent "point of departure" for calculations in the verification stage with certain though unclear anticipations for its results. The three-process theory formulated by [12] helps in identifying different types of insights. Selective combination takes place when someone suddenly puts together elements of the problem situation in a way that previously was not obvious to the individual. Selective encoding occurs when a person suddenly sees ... one or more features that previously have not been obvious [12]. Selective comparison occurs when a person suddenly discovers a nonobvious relationship between new and old information. Koestler's Act of Creation occupies the central illumination stage of those 4 or 5 stages pathway. His definition of bisociation, as a generalization of Aha! Moment and Eureka experience, allows to deepen our knowledge of the illumination stage into cognitive (and affective) aspects of the insight.

The second approach to creativity is via [13, 14] who were interested in the development of creative characteristic human attributes centered on divergent thinking. It is the creative product theory in distinction to the process-oriented Gestalt approach. In mathematics education its central qualities have been established to be fluency and flexibility of thinking assessed by speed and precision of thinking and the number of different solutions to the same problem. The third quality is originality measured by comparing individual solutions with those of the whole set of participants.

Originality of thought or action displayed through an Aha! Moment is the quality that joins Koestler's bisociation theory with Guilford's approach. However, this spontaneous originality is presented within the tension of automatization of a habit. In fact, Koestler sees in [2] the creative act of the insight as "an act of liberation—the defeat of habit by originality". He of course realizes that "habits are indispensable core of stability and ordered behavior, [yet] they also have a tendency to become mechanized and reduce man to state of conditions automata" (p. 96). This tension has a direct bearing on the mathematics classroom. Till recently learning of mathematics involved learning the procedures or codes and, actually, trying to make them automatic exactly as Koestler describes, because it increases the fluency. The habits are condensation of codes (procedures) learning. Recent curricular changes with the emphasis on problem solving have as a goal precisely lessening the grip of habits to create the conditions for creative solutions. If as Czarnocha in [15] suggests, creativity should be the foundation of learning, in particular of learning in mathematics, then we have a serious philosophical didactical problem to solve: What should be route of integration of creativity with the necessity of knowing and mastering procedures?

An important issue that arises in the debate between conceptual or relational instruction versus a procedural orientation is the nature of the concept that arises during a moment of insight. In social-constructivist theory, this issue can be understood within the context of Vygotsky's statement that unlike procedural knowledge, conceptual knowledge cannot be taught. Thus, it is the individual, guided by a mentor, who gives meaning to the cultural artifact being presented. In constructivist theory the concept-process is first interiorized and then reflected upon. This reflection upon the interiorized activity leads, through the emerging action scheme, to further structural development, which can be assessed with the help of Piaget and Garcia Triad. This structural development is called here the depth of knowledge (DoK) gained in the moment of insight.

3. The development of theoretical framework underlying the assessment of the depth of knowledge reached during the insight

3.1 The Triad of Piaget and Garcia

The Koestler-based definition of an Aha! Moment or Eureka experience insight of bisociation as a spontaneous leap of insight that connects unconnected frames of reference, matrices of thought is very informative. This suggests that the creativity of an Aha! Moment is in building new connections between different matrices of thought or more precisely between different schemas of thinking (that is, between networks of concepts and connections/relationships between them). Consequently, to understand a so defined creative process, we need to use a theory of schema development. The Triad of Piaget and Garcia developed in [7] turns out to be an excellent constructivist model of schema development whose three stages, -intra, -inter and -trans, allow us to characterize the cognitive levels of different Aha! Insights.

The -intra stage of the development of a concept in the mind of a learner consists of isolated instances, concrete manifestations of the concept as actions, rules or operations, whose performance requires an external help. The process of making connections between isolated instances of the concept means entering the -inter stage of the development. The transition between -intra and -inter case that is from isolated instances to connected instances of the developing concept is one of the sources of Aha! Moments because the transition between the two creates the necessary bisociative context. The -inter stage is rich in possibilities of different connections, from two concepts, single connection schema to multi-connections schemas, and each new connection made can be the effect of the creative insight.

The second important transition in the development of a concept is from the -inter stage connectivity to the -trans stage of generality or abstraction, finding the unifying principle or unifying structure of the concept. This transition is also good source for Aha! Moments. The thinking mechanism which propels the learner along these developmental pathways is called a reflective abstraction, that is, a learner's reflection on its own solution activity. Baker [16] has shown that a bisociative frame can be identified within the reflective abstraction process composed of two parts: projection and constructive generalization indicating a close relationship between bisociation and reflective abstraction.

Our aim is to create assessment of the depth of knowledge reached during the particular insight by the analysis of the nature of created schema of thinking. We take for that purpose numbers of connections made, the conceptual distance between the involved concepts and the quality of connections made. As the measure of that quality, we will take verbal or written description of discovered relationships.

3.2 Koestler/Prabhu theory of creativity

The presentation of the Koestler/Prabhu theory follows Redford, who in [17] formulates requirements for a theory in mathematics education more precisely by suggesting that "a theory can be seen as a way of producing understanding and ways of action based on:

- A system of basic principles, which includes implicit views and explicit statements which delineate the frontier of what will be the universe of discourse and the adopted research perspective.

- A methodology, M, that includes collection of data interpretation as supported by P.

- A set Q of paradigmatic research questions."

The phenomenon under investigation is moments of insight in the learning process: those leading to progress in understanding, or concept development. As teacher researchers our contention is that such moments occur both during social discourse, internalization, as well as during reflection on solution activity.

The basic principle underlying the Koestler/Prabhu theory of learning through creativity of Aha! Moment is the conviction that creativity is the cornerstone of that learning, which leads to progress in understanding. Aha! Moment as a *spontaneous leap of insight, which connects unconnected matrices of thought,* termed by Koestler *bisociation,* is the insight, which leads to progress in understanding through creation of the new conceptual connection, the new entities that is new (for the learner) knowledge.

The second principle of the theory is the ***bisociative frame:*** that is, those originally disconnected matrices of experience within which the Aha! Moment insight takes, or may take, place. Such moments of insight help transcend the learner or solver to a more universal plane within their community of peers and the mathematics classroom, producing positive affect. Since, as we show below, bisociative frame can be identified within different theories of learning, creativity can be seen as an irreducible component of learning. The bisociative frame is the possible creativity detector; it is the bisociative frame that is responsible for the coordination of creativity with different theories of learning, both on the practical level of the classroom as well as on the theoretical research level. The ease and the method with which a bisociative frame can be identified within the "host" theory is described by Czarnocha and Baker in ([4], Introduction). This process of identification has become the tool with the help of which we can indicate the areas of heightened creativity within the cognitive structure of the "host" theory; see examples in [18, 19]. The strength of this tool can be seen in the process of identification of creativity within Vygotsky's sociocultural theory within the process of internalization [20].

Koestler/Prabhu's theory of learning investigates the nature of creativity of Aha! Moment insight in several dimensions paradigmatic research horizons:

- In its relation to student affect and its relationship with conation, that is, with the possibility of satisfying deep human needs of the individual student;

- In its cognitive dimension with the hope of understanding the intrinsic structure of creativity as progress in understanding; it studies the nature and the development of the schema of thinking that is created during the Aha! Moment insight.

- In its theory of learning dimension, it studies the changes in understanding of mathematical concepts occasioned by Aha! Moments, the relationship between interiorization and internalization and the role of creativity in abstraction.

- In its networking theories' dimension, it investigates the processes of integration with different theories of learning during which bisociation can express itself in terms of the host theory.

The third principle of the theory is measurability of creativity, whose aim is to establish the depth of knowledge (DoK) reached during Aha! Moment insight; however, that needs to be done in a way that does not disturb creativity itself, or as [21] suggests, it is measured by internal variables. We use the theory of the Piaget and Garcia Triad [7], which allows us to assess the dynamic development of the new schema from the learner's or a scientist's description of the Aha! Moment insight coordinated with the researcher's view of the mathematical situation. Important disclaimer: Although we use a constructivist Piagetian concept here, we are not necessarily ascribing to the full scope of constructivist approach, although Koestler's point of view in his Act of Creation is not positioned very far. What is important, however, is that we can undertake an analysis of creativity within the sociocultural approach with the help of the same concept of the theory, the bisociative frame, which has been developed in the individual context. We see that Koestler/Prabhu theory has a consistent, independent relationships with both the Piagetian and the Vygotskian approaches. Investigation of that strange fact and its consequences is the research theme for a next investigation.

While the central paradigmatic question of the theory is two-fold

- identification of acts of creation in students thinking.
- identification of enhanced possibility of acts of creation within different theories of learning,

each of the horizons of the theory has its own paradigmatic questions, and the results of recent inquiries into the horizon of learning are presented in Section 4.

The methodology of the theory can be viewed through several angles, including the angle of teaching, of research and of assessment. Within teaching, we attempt to create mathematical situations that contain gaps in student understanding. For instance, we frame the questions, hints or assessment process using bisociative frames, which we suspect will be unconnected in student minds. Or sometimes we add a component that we suspect is needed for the student to gain an insight. In its research aspect, the aim of our methodology is to identify bisociative frames and hence to identify creative possibilities within the "host" theories of learning. Having done that, we have the possibility of using the methodology of a host theory to express the new forms of creativity within it. In its assessment aspect, we assess the change in the development of the relevant schema with the help of PG Triad.

An excellent example of such a coordination between bisociation and modes of reflective abstraction is presented by [18], who demonstrates the presence of a bisociative frame in two critical stages of his framework called "reflection on activity–effect relationship," the Ref* AER. The framework Ref*AER is the elaboration of a Piaget mechanism of reflective abstraction, within which Tzur [18] identifies two different stages of development, participatory and anticipatory. In the participatory stage, he finds six different categories of reflection, each equipped with a bisociative frame, suggesting the presence of sufficient cognitive conditions for facilitation of six, cognitively different Aha! Moments. That work immediately suggested to him a host of new paradigmatic research question. It is instructive how identification of bisociative frame within Ref*AER theory has broadened and deepened the theory itself.

Similar investigations have also been undertaken in [19] where identifying bisociation within different modes of attention allowed the introduction of dynamics within its structure. This bisociative dynamic results in a shift of attention from dyadic to triadic mode of attention, that is, from two components of the bisociative frame to both components together with the new relationship provided by the Aha! Moment.

This unusual capacity of a bisociative frame to be identified within different theories of learning and therefore in corresponding teaching practices provides an important opportunity to conceptually unify the constructivists and the sociocultural theories of learning. That pathway of inquiry has been initiated by Baker in [16] where he identified bisociative frames within the constructive generalization of Piaget, and in [20], he explored creativity within central sociocultural concepts of internalization and appropriation. One major effort of the theory is clarification of the relationship between the concept of interiorization arising in Piagetian theories and internalization of the sociocultural theory to which we devote Section 4.

4. The assessment of the depth of knowledge reached during the Aha! Moment insight

The definition of bisociation, whose cognitive content is to construct a connection between two unconnected frames of reference or matrices of thought, suggests the theory of the schema understood as the network of concepts to be the tool with the help of which we can trace the developmental aspect of the insight. DoK in this case will be the progress in understanding the relevant concept or the difference between student understanding of the mathematical situation before and after the insight. The Aha! Moments analyzed below are taken from the Collection of Aha! Moments in [4]. We established there three levels of DoK, namely mild, normal and strong.

4.1 Calculus AHA! Moment: mild

During my Calculus 1, the teacher gave us an example to solve: $\lim_{X \longrightarrow 0} \frac{\sqrt{1+X}}{X} - \frac{\sqrt{1-X}}{X} =$

1. I verify if the limit is defined when X approaching to 0. It is not.

2. I asked myself "how can I do and find a way for this limit can be defined?"

    I remember in my previews class math 150 when the teacher gave us a rational fraction to solve, he said that we must eliminate the radical in the denominator by multiplicated by the conjugate. But for this equation we do not have radical in the denominator but in the numerator.

    I'm a little bit struggling. What can I do?

3. I was looking at the limit and said to myself why not apply the same rule for the fraction when we have the radical in the denominator.

$$\lim_{X \longrightarrow 0} \frac{\sqrt{1+X}}{X} - \frac{\sqrt{1-X}}{X} = \lim_{x \to 0} \left(\frac{\left(\sqrt{1+x} - \sqrt{1-x}\right)\left(\sqrt{1+x} + \sqrt{1-x}\right)}{x\left(\sqrt{1+x} + \sqrt{1-x}\right)} \right) \tag{1}$$

$$= \lim_{x \to 0} \left(\frac{\left(\sqrt{1+x}\right)^2 - \left(\sqrt{1-x}\right)^2}{x\left(\sqrt{1+x} + \sqrt{1-x}\right)} \right) \tag{2}$$

$$= \lim_{x \to 0} \left(\frac{1+x-1+x}{x\left(\sqrt{1+x} + \sqrt{1-x}\right)} \right) \tag{3}$$

$$= \lim_{x \to 0} \left(\frac{1+x-1+x}{x\left(\sqrt{1+x} + \sqrt{1-x}\right)} \right) \tag{4}$$

$$= \lim_{x \to 0} \left(\frac{2x}{x\left(\sqrt{1+x} + \sqrt{1-x}\right)} \right) \tag{5}$$

$$= \lim_{x \to 0} \left(\frac{2}{\left(\sqrt{1+x} + \sqrt{1-x}\right)} \right) \tag{6}$$

$$= 1$$

4. Now the limit is defined. I can solve it and finish.

Analysis

The student reports the experience of an Aha! Moment in the Calculus class while solving a limit problem. The content of the student insight is the discovery of the analogy (hidden analogy) between the algebraic expression involved in the limit problem with the algebraic expression she learned in the previous algebra class topic on rationalization of algebraic fractions. The similarity suggests to her the method of conjugates used in the rationalization case as the method of solution for the limit problem.

Thus, this single-connection schema element is constructed from two very close yet separate (in the student's mind) representations of the pre-conjugated algebraic expression: one positioned in the denominator of the fraction, and the other in the numerator of the fraction. By applying conjugates method the student learned in the past for the rationalization purposes to the calculation of the limit she has understood (and verbalized) that the application of the method does not depend on the position of the pre-conjugated algebraic expression within a fraction.

We assess the DoK of this insight as Mild for two reasons: (1) the increase of understanding was of just one new connection and (2) the two unconnected initially frames of reference were conceptually very close to each other. This classification is based on the local nature of the search process leading to the moment of insight. This moment of insight takes place in second level of P&G Triad, as the moment of bisociation connects two existing matrices one introduced in calculus the other earlier in algebra. Finally, the concept formed was a new understanding of the limit process.

This moment of insight occurred as the student reflected on one scheme she was learning in calculus class M1 with another she had learned previously in college algebra course M2. Thus, it was mostly constructivist in nature, or a moment of untutored bisociation as Koestler would phrase it. However, it contains distinct relationships to insights within the internalization process, as the student begins with replicating a matrix M1 modeled in class: internalization.

4.2 Fir tree AHA! Moment: normal

Consider the following function that generates the geometric pattern of a reverse growing fir tree (**Table 1**):

Instruction: Draw and describe the Stage 5 of the patterns in terms of its shape and the number of unit squares.

ANSWERS of the student

Stage 1	 2 unit squares
Stage 2	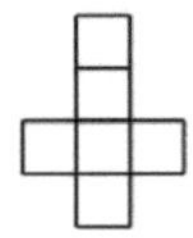 6 unit squares
Stage 3	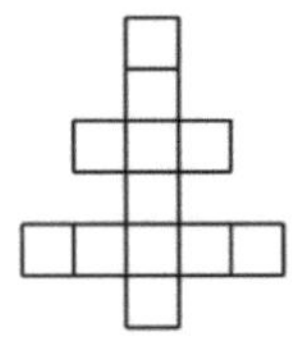 12 unit squares
Stage 4	

Table 1.
Stages of the fir tree assignment.

1. Stage 5's shape is bigger than the previous ones; it grows horizontally and follows the patterns of stage 4. It has 30 unit squares, meaning it increased by 10.

2. Describe how the patterns are growing.

   The pattern is growing by increasing 1 unit square to the left, 1 unit square to the left, one additional unit in the center and 1 unit down compared to stage 4.

3. How many unit squares are needed to build Stage 10 of the fir tree?

   Answer #3. It needs 110 unit squares. I developed this answer from [the discussion] next page

4. Given any stage **n** determine the close form of equation to determine the number of unit squares needed to build the tree.

   The formula is: $n(n + 1)$*.*

   Any stage number multiply by its following number equals the number of unit square, for instance: stage number: 3 multiplied by its following number meaning 4 equals 12 (unit squares.)

5. Your mate tells you that exactly 274 unit squares make up a fir tree. He is wrong. Explain to him why he is wrong.

Answer #5.
It is really wrong because:

Stage 16 = 16(16 + 1) = 272.
Stage 17 = 17(17 + 1) = 306.
The fir tree would never have 274 unit squares.

> *I had a tremendous Aha! Moment. I just realized that the formula I got from the patterns was a factorized expression and if I multiply it "n(n+1) = units square" I would have something like an algebraic expression exactly a trinomial expression that can be factorized as well, and it equals real numbers for example: n(n+1) = 12 n^2+n= 12 n^2+n-12=0 n^2+n-20=0 n^2+n-56=0 n^2+n-90 (n+4)(n-3)=0 (n+5)(n-4)=0 (n+8)(n-7)=0 (n+10)(n-9) And when it comes to n^2+n-274 it cannot be factorized.*

Analysis

The Aha! Moment took place after the student solved the assigned problem. From that solution, we know about the student's cognitive capabilities: She obtained the general formula using the variable n by generalization from the table of results for each term. She solved the question 5 by trial and error and demonstrated from the logic of these calculations that proposed result is incorrect. These are the components of the knowledge schema before the insight. The insight provided new conceptual solution to the same problem (**Table 2**).

It connected the factorized quadratic algebraic expression, a binomial with the corresponding quadratic trinomial by incorporating data of the problem. The second constructed connection is between quadratic trinomial and solutions of quadratic equation via its factorization. It is interesting that both connections are made through the process of factorization, and the concepts they connect are close to each other conceptually. However, the concept of quadratic equation is significantly further apart conceptually from factorization of trinomials, so much that the student does not recognize it within the mathematical situation. Yet the fact that student recognizes that constant terms in the binomials obtained through factorization are the solutions of unrecognized quadratic equation and have bearing upon the solution of the whole given problem indicates a larger conceptual distance within the second connection of the constructed schema. What also adds to the depth of the new schema is that from

Stage 1	1(1 + 1) = 2
Stage 2	2(2 + 1) = 6
Stage 3	3(3 + 1) = 12
Stage 4	4(4 + 1) = 20
Stage 5	5(5 + 1) = 30
Stage 6	6(6 + 1) = 42
Stage 7	7(7 + 1) = 56
Stage 8	8(8 + 1) = 72
Stage 9	9(9 + 1) = 90
Stage 10	10(10 + 1) = 110

Table 2.
Numerical relationships discovered at each stage of the fir tree Aha! Moment.

the language used it is clear the student (despite some holes in the overall schema of the situation) has a control over the "gestalt" of the schema—the student owns it—that is that it has been interiorized as well as internalized, with evidence for both contained in the student's description. The evidence of internalization is in the ability to explain the logical connections verbally, and interiorization is evidenced by the ability to provide the second solution, hence independent action upon the problem. We assess this progress of understanding at the second level of schema construction and name it as normal.

4.3 The domain AHA! Moment: strong

The problem starts with the function f(x)= $\sqrt{X+3}$. The teacher asked the students during the review: "Can all real values of x be used for the domain of the function f(x) = $\sqrt{X+3}$?"

1. Student (S): No, negative x's cannot be used.

2. Teacher (T): How about x = -5?

3. S: No good.

4. T: How about x = -4?

5. S: No good either.

6. T: How about x = -3?

   Student, after a minute of thought:

7. S: It works here.

8. T: How about x = -2?

9. S: It works here too.

   A moment later the student adds:

10. S: Those x's which are smaller than -3 cannot be used here.

11. T: How about g(x) = f(x)= $\sqrt{X+1}$?

    Student, after a minute of thought:

12. S: "Smaller than 1 cannot be used."

13. T: In that case, how about h(x) = f(x)= $\sqrt{X-a}$?

14. S: Smaller than a cannot be used.

DOI: http://dx.doi.org/10.5772/intechopen.110694

Analysis

This Aha! Moment clarifies student misconception concerning the domain of the $\sqrt{X+3}$. The original and habitual student answer in line 1 represents student misconception. The intent of the teacher's question was to direct the student's attention to the contradiction of her answer with the situation at hand, what led the student into cognitive conflict. The Aha! Moment takes place as a resolution of that conflict. The student's short reflection on the previous verbal interaction results in the correct reorganization of her approach.

Note that student's insight engages in her the domain of the function f(x)= $\sqrt{X+3}$ and order on the real number line. The bisociative frame of this bisociation were the axes of the cognitive conflict: the habit of well-established misconception and the data brought to the student's attention by focusing her attention on the relevant details. The conceptual distance between these components is not very large – just the change of the parameter but to traverse it the student had to engage her schema of addition for integers, which makes student thinking a bit more complex and places this insight on the second level.

Her language is straightforward, does not convey any doubts and places understanding on Normal level of DoK. However, follow-up questions of the instructor reveal deeper levels of understanding. For instance, the student can easily transfer her understanding to a different related example (line 11). More important, she can abstract and generalize the answer for arbitrary parameter a (line 14). This is the second-order reflection on the "family type" expressions, which leads to abstraction with generalization of the -trans stage. That indicates significantly larger conceptual distance between the components of the bisociative frame and places the Domain Aha! Moment on the third Strong level.

5. Koestler theory of creativity, Aha! Moment and learning

For Koestler creativity and indeed learning what is subjectively new for an individual, which he refers to as progress in understanding, takes place through the synthesis of two distinct and previously incompatible frames of reference. He refers to these as matrices, each of which has its own rules of the game or codes that govern appropriate activity.

What is a matrix of thought, or a frame of reference?

> *The matrix is the pattern before you, representing the ensemble of permissible moves. The code which governs the matrix can be put into simple mathematical equations… or it can be expressed in words. The code is the fixed invariable factor in a skill or habit, the matrix is the variable part. The two words do not refer to different entities, they refer to different aspects of the same activity.* (Koestler ([2], p. 40)).

Referring to individual moments of insight within the learning process, Koestler notes that "*Minor, subjective bisociative processes do occur on all levels and are the main vehicle for untutored learning*"(p. 658). This raises another important question addressed in [16] and in [20], how does one describe interiorization in Koestler bisociative frame? The more general question of how Piaget's notion of reflective abstraction fits into Koestler's bisociative frame is discussed in [22]. Since constructivist research methodology can be described as minimally guided, it certainly classifies as untutored and therefore should be bisociative in nature. Can the same be said of internalization?

Our analysis of the genesis or birth of a code during a moment of insight, whether during interiorization, internalization or bisociation, is conducted through the lens of three defining characteristics of such moments. (We will use here Koestler's term "blocked situation" as one in which routine matrices fail to accomplish the desired goal: to solve the problem). The first is the search process in a blocked situation. The second is the connection realized during the moment of insight between the blocked situation and a matrix that provides conceptual reasoning that allows for resolution of the blocked features. The third is the novel concept and process based on the concept that acts on the formerly blocked features to obtain the goal.

5.1 Discovery of a hidden analogy, interiorization and internalization

Koestler describes a blocked situation as one in which routine matrices fail to accomplish the desired goal. Constructivists use the term non-assimilatory situation. Koestler describes the process of trying to resolve a blocked situation by searching for a connection to an analogous matrix as the discovery of a hidden analogy, as a search for something that is unknown:

> *[T]he subject looks for a clue, the nature of which he does not know, except that it should be a 'clue'... a link to a type of problem familiar to him ... [H]e must try out one frame after another ... until he finds the frame into which it fits, ... an analogy with past experience and allows him to come to grips with it.* (pp. 653-654).

This search process in such a blocked situation to find a connection to an appropriate matrix-scheme in our repertoire can itself be considered a matrix-schema Mo. The search process follows in its general outlines the steps of [8, 9]. We may consider the code to contain some or all of the following guiding principles. First, identify relevant features of a problem situation that are blocked. Second, review our toolbox or collection of matrices that are even remotely associated with these features. Finally, select and proceed-verify the one most appropriate. If this fails, Koestler notes that mathematicians recommend to sit tight and wait for inspiration.

We note that contemporary instructional methodology of teachers encourages students to seek external assistance during the process of these transitions. As a result, the search process from constructivist pedagogy turns to internalization. This search process is highly subjective. Individuals vary in their ability to discriminate or identify what objects are relevant and abstract the conceptual relationship between these objects and appropriate activity. There is also a wide variety of motivation. Patience on the part of individuals during this search and finally even the motivation to seek assistance varies. All these factors impact the success or failure of this process.

5.2 Interiorization

Interiorization occurs during a child's transition from empirical-spontaneous to abstract reasoning, and thus, M_1 is an intuitive matrix and hence limited to situations where input present low cognitive demand. When a situation involves objects with more structure or cognitive load, the subject becomes perturbed, their attention has two foci—it is so-called dyadic attention. One is their Mo search to understand the structural objects that cannot be assimilated, and the other is that part of the M_1 matrix that acted upon the analogous spontaneous concepts. In a moment of bisociative insight between the search matrix and their intuitive matrix when the shift of attention shifts from dyadic to triadic attention of seeing the whole [19], the subject abstracts the spontaneous concepts.

This allows the formerly intuitive reasoning to become conceptual-based reasoning, and a pseudocode is born, through the bisociation of Mo and M_1. This particular description of interiorization in creativity theory is known as selective encoding. In this, the subject's attention is on a relevant matrix-scheme or tool that they previously could use only in a limited context and their insight allows them to use it more fully. Thus, interiorization as described consists of bisociation through selecting an object(s) that cannot be assimilated, search to connect them with the spontaneous concepts of a person's intuitive scheme, and the result is scientific concepts simultaneous with a process or conscious scheme.

5.3 Internalization and interiorization: the search matrix Mo

The distinction between interiorization and internalization as it relates to moments of insight in the learning process begins with the nature of the search matrix Mo. In Koestler's description of searching for a hidden analogy, the search matrix Mo is focused on the subject's collection of even remotely related schemes in the hope of finding an unknown connection. The second tenet of constructivism is that all prerequisite knowledge should be present in a learning situation.

Thus, the search process is typically on solution activity determined by an existing scheme (M_1) that is deemed appropriate and yet remains insufficient to resolve the non-assimilatory situation. As such, interiorization involves selective comparisons between the non-assimilatory objects in Mo and an intuitive M_1 leading to an abstraction of the underlying conceptual-invariant relationship. Reflective abstraction is the name given by Piaget to the mechanism of accommodation that includes interiorization. It has two steps. In the first step, an appropriate M_1 is projected into the search matrix Mo for a non-assimilatory situation. The second constructive generalization step involves perturbation between and reflection upon these two matrices. The creativity of Aha! Moment arises during the search within unconscious or semi-conscious Incubation Gestalt stage through the shift of attention from the separate matrices to the whole revealing the new structure.

Social constructivists use the term reflective thinking, instead of reflective abstraction, to include any form of conscious reflection on collective discourse, for example, shared activities, including observation of, or communication with, an exterior source of knowledge. Thus, the search process Mo in internalization includes co-creation of knowledge in all its forms, making it ideal for analysis of classroom discourse. Social constructivists use the term socially mediated activity to describe solution activity that involves assistance through any form of cultural artifacts, or guided communication, for example, written text, pre-recorded videos, other internet search sources, peer-mentor, or teacher etc. As students attempt to internalize material presented in the classroom, their primary motive is to understand or assimilate externally directed activity. Thus, they may have a motive but not realize the problem goal.

5.4 Moments of insight: connection

The moment of insight has a transcendent nature as a student is led to this connection by intuition not reason. As a result, in Koestler's view, pedagogy should be (re)structured to support such moments of insight into the guided discovery as math and science cannot be appreciated outside such experiences.

In the constructivist frame, during interiorization the moment of insight occurs as the individual struggles to employ an intuitive situation-dependent scheme M_1 in a non-assimilatory situation, that is due to the presence of Mo objects beyond the capacity of spontaneous thought. The perturbation between Mo and M_1 is resolved in a moment of insight or, shift of attention which abstracts the formally intuitive reasoning. This experience transcends the reasoning from the situation as it becomes an abstracted activity–effect relationships. This moment of insight or abstraction allows the individual to have conscious control over the process and thus to act independently. Thus, for constructivists, Mo contains an existing M_1 scheme, Situation/Activity/Effect (S-A-E) triad, which in a moment of insight is transformed into an abstracted Activity/Effect (A-E) dyad. That is one not dependent upon the situation. The pedagogy is guided discovery to promote reflection and abstraction upon such an M_1. In Koestler's search for a hidden analogy, the focus is on the formation of a S-A-E triad. During internalization, as in Koestler's frame, there is typically no existing S-A-E link within the search matrix Mo.

However, the pedagogy of internalization is not centered on guided discovery but instead on the mentoring of students as they accommodate new situations at the upper limit of their zone of proximal development. That is the upper limit of what they can realize with assistance. So, where is their own individual creativity? It must be beyond the assistance: beyond the upper limit of their individual ZPD. Where is the bisociative frame?

5.5 Creativity as deviation during the process of appropriation

Here, we can use an important suggestion provided by [23] who informs that internalization encounters difficulties in the proper description of the concept formation, also from the sociocultural points view. To address these problems, we need to introduce the concept of appropriation. The term appropriation is often used to analyze learning within social discourse; it has its origins in the work of the Russian psychologist M.M. Bahktin to understand how children learn language.

> *The word in language is half someone else's. It becomes "one's own" only when the speaker populates it with his own intention, his own accent, when he appropriates the work, adapting it to his own semantic and expressive intention.* [24]

When analyzed in sociocultural terms as the interaction between learners in the learning community, appropriation facilitates the development of concepts of individual learners. Let me continue here with the words of [23]:

> *"Where is creativity related to the appropriation process? This is its own deviation. Table 2 lists some of the possibilities of a wide variety of deviations. This includes the gap between the concepts of the learners interacting with each other and between the concepts of the learning community as seen by each learner. The cause of these deviations lies in the learners' historical and cultural constraints. By interacting according to*

> *these deviations, each learner may misunderstand what he/she is talking about. Simultaneously, the learner may create a new concept not included in the speaker's concept. In other words, the deviation of appropriation can be the source of creativity that leads to new ideas that the speaker did not intend."*

In other words, creativity in sociocultural theory originates in the differences between the learner's and the community's matrices of thought owing to different historical and cultural constraints. Hence, we have recovered the bisociative frame between the two matrices of thought at exactly the unique site of creativity within sociocultural theory. If one of two matrices is that of the community and another of the learner, it might be natural to call creativity a deviation from the community's point of view. It is interesting to trace out how that particular process of appropriation of the concept interacts with the relevant inner spontaneous concepts of the learner. We leave that subject as an open research question: how do appropriation and interiorization interact?

6. Conclusions

This work presented a new Koestler/Prabhu theory of creativity within learning and provided some investigations suggested by the theory. One of them was to investigate whether the bisociation frame as the central component of the theory of the creativity of Aha! Moment can be identified within the constructivist and the sociocultural approaches to learning. Our discussion in the first section where Koestler's theory is discussed indicates that the term and the concept of bisociation derive from the observation and analysis of individual insights. On the other hand, sociocultural approach places the emphasis on learning within collaborative teams rather than on the individual activity. Consequently, to find the space within that theory where social creativity can be understood with the help of bisociation as the theory of the individual insight is very important especially from the point of view of classroom teachers of mathematics, where both individual as well as collaborative learning are taking place.

Section 1 examined bisociation theory, interweaving it with the constructivist framework and followed by the preliminary discussion of the nature of creativity of Aha! Moment/Eureka experience. In Section 2, we presented Koestler/Prabhu's theory of creativity in mathematics following [17] who formulated the requirements for a theory in mathematics education. Koestler/Prabhu's theory stands on three principles: (1) the definition of an Aha! Moment as the object of our investigations, (2) the bisociative frame and (3) the measurability of creativity as the assessment of DoK reached during the insight.

At present, the theory has established four horizons of its investigations:

- A relationship of the insight with the affect and conation.
- The cognitive dimension, which studies the nature and the development of the schema of thinking created during the insight. We see this dimension of research in Section 3 where the cognitive mechanism of thinking called PG Triad is used for the analysis of DoK.
- In its learning horizon, we study changes in understanding of concepts occasioned by Aha! Moments. In particular, Section 4 explains the central

investigation of the relationship between interiorization and internalization in the context of bisociative creativity. As interiorization is a constructivist concept while internalization is a Vygotskian, sociocultural concept, this section has a bearing upon the goal of unifying individual and sociocultural approaches to creativity in mathematics education. In the section, we show the method of our approach and note its limitation due to it being too rigidly bound with the concept of ZPD as a framework for mentor–mentee social interaction. We discovered that the much more useful concept to investigate bisociative creativity within that approach is the concept of appropriation, and in particular that of deviation during the process of appropriation. As we indicated there within deviation of the concept of the individual learner from that of the community, the bisociative frame can be identified as two matrices of thought, which are different due to their historical and cultural constraints.

- In the networking theories dimension, we investigate processes of integration of Koestler/Prabhu theory with theories of learning what enables bisociativity express itself in terms of host theories. We have touched upon interaction of bisociativity with the theory of reflective abstraction of Piaget, theory of attention of John Mason and participation/anticipation approach of Tzur. Intimate connections of bisociative creativity with learning theories enables us to introduce creativity into different pedagogies of learning.

- Section 3 presented the method of DoK assessment on three different types of Aha! Insights, taken from the Collection in [4]. Section 4 explored our investigations leading to the identification of the bisociative frame exactly where it should be according to sociocultural approach, in the deviation within the process of appropriation.

Author details

Bronislaw Czarnocha* and William Baker
Hostos Community College, CUNY, NYC, USA

*Address all correspondence to: bronisuavec2@gmail.com

References

[1] Prabhu V. The creative Learning environment. Chapter 2.4. In: Czarnocha B, Baker W, Dias O, Prabhu V, editors. The Creative Enterprise of Mathematics Teaching-Research. Rotterdam, Boston, Taipei: Sense Publishers. pp. 107-126

[2] Koestler A. The Act of Creation. London: Hutchinson & Co.; 1964

[3] Czarnocha B, Baker W, Dias O, Prabhu V, editors. The Creative Enterprise of Mathematics Teaching-Research. Rotterdam, Boston, Taipei: Sense Publishers

[4] Czarnocha B, Baker W, editors. Creativity of an Aha! Moment and Mathematics Education. Leiden Boston: Brill|Sense; 2021

[5] Czarnocha B. Towards a philosophy of creativity in mathematics education. In: Bicudo M, Czarnocha B, Marciniak M, Rose M, editors. Ongoing Advancements in Philosophy of Mathematics Education. Cham, Switzerland: Springer; 2024 [In press]

[6] Czarnocha B. Facilitation of Aha! Moments in mathematics classroom. In: Poster Presentation, PME 41, Singapore, July 2017, Annual Meeting of PME 41. Singapore: PME

[7] Piaget J, Garcia R. Psychogenesis and the History of Science. New York, NY: Columbia University Press; 1987

[8] Wallas G. The Art of Thought. San Diego, California: Harcourt Brace & Co.; 1926

[9] Hadamard J. Psychology of Invention in the Mathematical Field. Princeton, NJ: Princeton University Press; 1945

[10] Poincare H. Science and the Method. New York: Dover; 1914

[11] Sadler-Smith E. Wallas' four stage model of creative process: More than meets the eye. Creativity Research Journal. 2015;**27**(4):342-352

[12] Davidson J. The suddenness of insight. In: Steinberg R, Davidson J, editors. The Nature of Insight. Cambridge, MA: MIT Press; 1996

[13] Guilford JP. The Nature of Human Intelligence. New York: McGraw Hill; 1967

[14] Torrance EP. The Torrance Tests of Creative Thinking-Norms-Technical Manual Research Edition-Verbal Tests, Forms A and B-Figural Tests, Forms A and B. Princeton, NJ: Personnel Press; 1966

[15] Czarnocha B, Baker W. Creativity and bisociation. In Singer FM, Toader F, Voica C, editors. The 9th Mathematical Creativity and Giftedness International Conference. Sinaia, Romania. 2015. pp. 80-85

[16] Baker W. Koestler's theory as a foundation for problem-solving. In: Czarnocha B, Baker W, Dias O, Prabhu V, editors. The Creative Enterprise of Mathematics Teaching-Research. Rotterdam, Boston, Taipei: Sense Publishers; 2016. pp. 267-286

[17] Redford L. Connecting theories in mathematics education. ZDM: Mathematics Education. 2008;**40**:317-327

[18] Tzur R. Two state changes in anticipation: Cognitive sources of Aha! Moments. In: Czarnocha B, Baker W, editors. Creativity of an Aha! Moments

and Mathematics Education. Leiden Boston: Brill|Sense; 2021. pp. 321-344

[19] Mason J, Czarnocha B. Aha! Moments, bisociation and multifocal attention. In: Czarnocha, Baker, editors. Creativity of an Aha! Moment and Mathematics Education. Leiden Boston: Brill|Sense; 2021. pp. 345-364

[20] Baker W. A framework for creative insight within internalization of mathematics. In: Bicudo M, Czarnocha B, Marciniak M, Rosa M, editors. Ongoing Advancement in the Philosophy of Mathematics Education. Cham, Switzerland: Springer; 2024

[21] Morruzzi C. An account of natural and artificial creativity. European Journal for Philosophy of Science. 2021; **11**:1

[22] Baker W. Bisociation, creativity and interiorization. In: Czarnocha B, Baker W, editors. Creativity of Aha! Moment and Mathematics Education. Leiden Boston: Brill|Sense; 2021. pp. 301-320

[23] Matsushima M. A reconsideration of appropriation from socio-cultural perspective. In: Bicudo M, Czarnocha B, Marciniak M, Rosa M, editors. Ongoing Advancement in the Philosophy of Mathematics Education. Cham, Switzerland: Springer; 2024

[24] Bakhtin MM. The Dialogic Imagination: Four Essays. 9th ed. Vol. 1. Austin, Texas: University of Texas Press; 1994. p. 293

Section 2

Creativity and Innovation: An Eternal Bond

Chapter 4

Creativity Impediments and Solutions

Josam Nandwa Musambayi

Abstract

The chapter covers 'Creativity Impediments and Solutions' in organizations in Kenya. This chapter addresses what impedes creativity in organizations and yet creativity informs the entrepreneurial mind. Creativity enables one to move ideas from theory to reality. This chapter covers the introduction part explaining innovation, entrepreneurship and creativity. These are considered to be at the very core of entrepreneurship. The second part covers the barriers to creativity and the approaches the researcher employed in gathering data. It goes further by highlighting the measures of enhancing creativity before making conclusions. This part goes into the depth of the matter as an integral part of this book chapter. In the end, the chapter looks at the suggestions that have been identified as solutions of addressing this creativity gap and impediments. This is the very section that also captures the author's imagination of the subject matter. It is hoped that this chapter will go a long way in enabling policymakers, scholars and entrepreneurs to effectively utilize creativity to support innovation and entrepreneurship development. This is an outcome of what the researcher has gathered from engagement with university scholars, industry players, experts in the field and other stakeholders over a span of time.

Keywords: creativity, impediments, work environment, solutions, organization

1. Introduction

Entrepreneurship considered to be one of the engines that drives economies and societal activities is a process that also needs other inputs. This means it requires others ingredients other than the human mind to thrive. One such input is creativity which, other than being an idea, it is a process too. It is important to briefly highlight the three closely linked concepts before the actual topic is tackled. As Bao [1] opines, one of the most fascinating descriptions of creativity is that ability being able to come about with not only new, but also innovative ideas which can be a valuable asset in many fields. In this case, the various fields can include business, education and research, as well as in the matters industrialization and in other societal spheres. The scholars further describe the individuals in this discipline as being extraordinarily creative. As it were, this can easily lead to innovation. The same creative people are termed as persons who go around asking, 'Why not?' What this means is that individuals of these stature are able to see not only new combinations, or new potentials, but also new opportunities and challenges particularly where others see difficulties or limitations [2].

Innovation for instance, simply put, is the value addition to what scholars, researchers or firms could be doing. It can be about innovation of a product or service, a business or a technology-a system or method of production. In essence, innovation is improvement which revolves around 'doing something'. This is a very important concept in the field of entrepreneurship development across the globe. It is even argued that that entrepreneurship cannot be considered complete without innovation. This cannot be put any better than what Schumpeter theorized as the newness of products or services, identification of new markets or customers, seeking new sources of raw materials or supplies to the firm, adopting new and efficient methods or systems of production as well as embracing the new forms of businesses today [1]. Innovation has become necessary in society today to the extent that every basic aspect of human life is controlled by various aspects of innovation. But the bottom line is in the doing which relates to performance of improvement of undertaking things [3]. The matters innovation in the present age is considered to be very essential particularly when it comes to manufacturing. One of the reasons is that the end users of products are keen on the quality of the product they buy. The quality of the product is influenced to a large extent by the process it undergoes in transformation of the raw material to the finished product. This is how innovative manufacturing comes into play. In the emerging global economies, innovation is also taking centre stage as value addition activity which ensures there is improvement to businesses, technology or systems of production as well as on product or services. The process further improves quality and competitiveness in the market place.

Concerning entrepreneurship, scholars consider it to be a process of engaging into certain activities within a setup with specific intentions arising from the entrepreneur's mind or concept. This process incorporates the person into a process that results in an entity referred to as an enterprise. The three important elements of entrepreneurship are the person (entrepreneur), the process (entrepreneurship) and the product (enterprise). Entrepreneurship is informed by an existing need in the society known as opportunity. The opportunity must be responded to by the entrepreneurs coming up with appropriate ideas as solution to the identified need or problem in a societal setup. Globally, entrepreneurship is considered to be the engine behind various economic activities. In essence, entrepreneurship is running many economies and is helping those in government to solve their problems of job opportunity creation; they are also acting as the sources from which governments raise their incomes in form of taxes to fund other activities in the countries. In brief, entrepreneurship through the entrepreneurs contributes in bringing about social, technological and economic changes in nations. Therefore, it goes without saying that those in authorities are doing everything within their power to support the establishment and development of entrepreneurship [4]. But in order to support this venture, a lot has to go into thinking and thinking creatively. One of the reasons for this is that viable ideas are likely to emerge from creative thinking and such ideas are the same ones that end up catalyzing entrepreneurial activities. As earlier opined by some scholars, entrepreneurship must be conducted by having change of mindset. This is what was alluded to as 'disturbing the status quo'. This process requires that an entrepreneur does not only become passionate about business, or good at moderate risk taking, but must also be creative in all the ways, particularly in creative thinking and in the way of conducting the activities.

Creativity on its part is one of the most important aspects of entrepreneurship. What this means is that new ideas must be well thought out as essentials to the process of entrepreneurship. Creativity is in the 'thinking' and thinking new ideas which can later be turned into innovation for the firm and entrepreneurship development. Creative thinking requires people, an environment and support from not only the entrepreneur, but also the authorities and policymakers [5]. It has been revealed there are certain efforts that go into creative thinking, some of which conscious and others unconscious. It therefore means that deliberate measures need to be taken towards this end [4]. What is also becoming apparent is that creativity is absolutely necessary for firms' survival in the world of entrepreneurship.

Firms need to involve their teams in creativity processes by assuring the members that success of every individual person contributes to that of the entire organization. It should be pointed out that failure in any of the steps is to be taken positively and should be considered as a point to learn.

1.1 Creativity: a necessity for survival

Firms, corporate entities as well as societies need creativity, not only for performance, but for survival and sustainability. Creativity informs entrepreneurial mindset and professional performance. If one wants to truly establish a great business, then they must apply creativity to such businesses. The entrepreneurs get paid to solve problems, and hence, they must be very creative. This is also true because creativity enables firms and societies to develop through certain modifications by taking into account the things and ideas that evolve over time. This notion is critical for flexibility of appreciating change, an essential aspect for societies and firms that need to adjust quickly to survive and flourish into unforeseeable future. Sankaran [6] posits that creativity in organizations is absolutely essential. The implication is that such firms cannot survive without embracing creativity. One of the responsibilities of the top management is essentially to ensure that creativity is not only embraced in the organization, but also practised for the survival of the firm.

1.2 Creative thinking

Creative thinking, which is the ability to come up with unique and original solutions, is required and is very essential to establishment of firms and entrepreneurial culture in any country. It is equally necessary in organizations. Sometimes it is referred to as creative problem-solving; hence, creative thinking is a valuable and marketable soft skill in a wide variety of businesses [7]. Creative thinking being a process of considering situations, from innovative perspectives to developing an open-minded mindset, is a powerful concept in the discipline of entrepreneurship development. By expending creative thinking process, one is able to generate new ideas, by interpreting circumstances, identifying themes and designing alternatives to existing approaches of operating and performance. It is the capability of considering something in a new or improved way [7]. Creative thinking can also involve a close examination of progressiveness, problem-solving, organization, as well as communication. Various employers consider creative thinkers highly, and therefore, the workers should consider highlighting their creative thinking skills when called upon time and again. The process of creative thinking acts like a seed-bed to innovation and entrepreneurship [8].

1.3 Work environment

Creativity and innovation as concepts, just like products or services, need certain environment in which they can thrive. Since these are thoughts which result in emergence of creative ideas, they are formed in the minds of individuals. Situations to make this flourish must be encouraged, team spirit should also be fostered, and employees as members of the organization have to be involved. This involvement starts with awareness creation, identification of talents and individuals as well as incentives provisions as a way of ensuring creativity and innovation emerge and grow. The best place for this to happen seems to be the workplace as the right environment [9].

1.4 Solutions

One of the intents of entrepreneurship is to solicit for resources and individuals to come up with solutions for societal problems. In view of the ever-increasing demand by consumers of services, experiences and products in the society, formulating solutions is equally necessary. This will suffice for customers, organizations as well as the governments of the day. But the aspects of creativity and innovation play a major role to the coming up with solutions [10].

1.5 Organizations

It is emerging that various organizations are beginning to embrace creativity, innovation and entrepreneurship as strategic measures for sustainability. Those that lead the pack are business organizations and the research institutions. But closely following are the institutions of higher learning and to some extent the government institutions. It is thought that if more stakeholders are involved in this venture, then it is possible that the outcome of research and effects and impact of creativity and innovation in particular will be felt large and wide. This should trickle down to individual entrepreneurs and groups of corporate organizations. When such a scenario is witnessed, then it is possible that seeds of industrialization can start to germinate [11].

2. Barriers to creativity

Barriers to creativity is also be referred to as the creativity gap, and it is becoming an area of concern to the establishment of firms, performance of the same and ultimately the development of entrepreneurship in countries. Some of the impediments to creativity as the study extracted range from attitudinal to training and even aggravated by the insufficiency of budgetary allocation. The creativity gap is also causing a number of scholars to initiate research studies in order to interrogate the said gap. The barriers seem to be at specific levels of the organizations. But before engaging in the actual investigation of the creativity gap, the research approaches were briefly highlighted for the understanding of the study in question [10].

2.1 Approaches

The data collection and analysis of the same in this study has taken several months starting sometime in February 2020 to date. This was the time when the first case of COVID-19 had just been reported in Kenya.

2.1.1 University class engagement and discussions

The researcher, who is an instructor at the university, engaged other scholars at universities on this subject. The scholars were sampled from two public universities, one private university and a TVET. Some of the respondents were not available for face-to-face interview and had to be contacted using other channels. Their responses were captured and analysed to give an insight in the subject under study.

Sampling was systematically done for purposive reasons on desired outcomes as indicted in **Table 1**.

2.2 Analysis of information from stakeholder organizations

Analysis of information from stakeholder organizations such as WIPO-PRV Training, KIPI, KECOBO, KeNIA, NACOSTI.

These organizations are the major stakeholders in matters of innovation, creativity and intellectual property management. The information concerning creativity was obtained from their publications and Websites. The information related to training, regulation and registration, funding among others. The sought data obtained was analysed, and important conclusions were drawn from the same.

2.2.1 A meeting and interviews with industry players

A meeting and interviews with industry players who included representatives from commercial banks, The Kenya Seed Company Ltd., Knowledge Transfer Network (KTN), Eldo Bub, and County Government of Nandi. These organizations were represented by high-ranking officials who gave their views on creativity and the possible industry and university collaborations on the subject at hand.

The officials were purposively invited on the account that they represented their organizations on this important topic. This is shown in **Table 2**.

2.2.2 Input from experts

Expert opinion was sought from among others, Technology Transfer Manager [Moi University], intellectual property [IP] official from the University of Eldoret

University	Category	Tutor	Staff	Year 2	Year 3	Year 4	Total.
Moi	Public	2	3	5	7	10	27
U.o.E	Public	2	3	5	7	10	27
Mku	Private	2	3	5	7	10	27
TVET	Public	2	3	5	7	10	27

Table 1.
Respondents who completed the questionnaires on creativity.

Company	Category	Staff	Cadre
Absa	**Bank**	2	**Mgrs.**
Kenya Seed	Public	1	Mgr.
KTN	Private	1	Rep
EldoHub	Public	1	C.E.O
Nandi County	Government	1	D-Gov

Table 2.
Respondents who were interviewed during the meeting.

[U.o.E.] and Eldo-Hub Head. These three individuals who are experts in the area of innovation, research and intellectual property were extensively interviewed on the subject of creativity. They also made suggestions on the way forward of fostering creativity. From those interview schedules and proceedings, very essential information was obtained, and critical insights and conclusions were made in connection with creativity. This became the basis for making findings of this study for the purposes of disseminating the results.

2.2.3 Output from innovation workshops: Public lectures

As a way of creating awareness on innovation, the university arranged innovation week in the year 2021 and 2022.Other than the exhibitions that were displayed, there were engagements among the innovators and participants, and the culmination of this was public lectures that were given by scholars such as Dr. Jesper Vasell, the Deputy Director of NACOSTI in Kenya, the C.E.O of KeNIA among others. In this endeavor, the then Director of Innovation & Research at the University of Eldoret, Prof. Vincent Sudoi was very instrumental. The proceedings from these sessions of the Innovation week illuminated so much on the subject of creativity and greatly assisted in extracting some of the issues that had been identified as challenges.

2.3 Impediments to creativity

One of the ways of eliminating barriers to creativity is removing things that clutter the mindset. Impediments, barriers and gap are terms that were interchangeably used. In this case, mindset issues that contributed to barriers or impediments included the thinking state, which is the concern for certainty, and formed opinions about a certain aspect of concept. This limits creative thinking. Another issue is security hunting—it is the concern for risk. It makes one become risk averse. We also have stereotyping which is a case that keeps abstracting reality. The probability thinking, on the other hand, seeks predictable results, which ultimately impedes creativity.

2.3.1 Fear

Fear is one of the biggest reasons that make organizations as well as individuals not to achieve their full potential in creativity. They have the fear of failure in their endeavors. The fact that there is no assurance that the new idea will work makes the creative people and firms to develop fear. Fear consequently slows the creative thinking process. As it is, creative thinking process is a psychological or intuitional one. When fear is recorded, it hampers or inhibits thinking. However, coming to terms with the fact that failing is an important part of learning; when handled well, it can lead to development and progress [1].

2.3.2 Leadership issues

Studies show and scholars agree that issues of innovation and creativity can be well spearheaded from the top leadership of organizations. The top executives and directors of various firms are yet to strongly come out and support the strategic importance of innovation and creativity as key elements of corporate performance. It would appear that currently corporates are yet to take leadership on this subject in order to foster performance and development. It is thought that if top leadership of organizations changes its attitude and approach to creativity, then the performance of various firms can also change. But it means the top managers must now change and lead innovation and creativity from the front in their respective entities. They also need to start nurturing upcoming innovators and creative thinkers [12].

2.3.3 Deficient resources and capacity

Investing in creative activities and in innovations can be a very expensive affair, and the same requires massive finances and human capital. Equally needed are systems, training infrastructure and the experts. All these require huge amounts of funding which is insufficient. This is a big impediment for this creativity which is a necessary endeavor. However, firms, institutions and countries must take deliberate decisions and set budgetary provisions for this purpose. They can also employ strategies such as networking and collaborations in a bid to foster the creativity task.

2.3.4 Lack of necessary alliances

Internal budgeting and structures may not assist alliances between teams. It is reported that many years later, organizations do not appear to have embraced the required collaborations to have things changed. It also seems that various barriers are still existing. These barriers could be similar across organizations and among individuals. They still impede creativity significantly. Lacks of strategic alliances among other factors still inhibit the ability to think creatively and hence innovate. Alliances must be encouraged to foster creativity and help in the achievement of the desired results. Creative collaborations are also about bringing people together, viewing challenges from different viewpoints and then linking that knowledge and understanding together, to create a solution. This must be a solution that can then be strategically planned and implemented effectively [13].

2.4 How to enhance creativity

Various scholars in their studies have attempted to find solutions to some of the impediments that have been identified so far. Although the impediments identified earlier are weighty, they are only a representation of the myriad of other challenges. Therefore, the suggestions here are just a few ways of enhancing creativity. It means a lot more can still be done through initiating further research [14].

2.4.1 The creative process

According to Corazza and Agnoli [13], whether one calls it creativity, creative thinking or just creative ideas, there must be a process involved in it. This process needs to be meticulous and can also be learned because it involves individuals, firms and systems. Above all, it must be supported financially, morally and by establishing the right systems.

What is fascinating is that creative thinking process is largely a subconscious process. Due to this fact, the individuals must be targeted because this process takes place intuitively.

2.4.2 Techniques for improving the creative process

The findings indicate that there is the need of improving the creative process. If this process is improved, then the ideas that can be obtained as a consequence can go a long way in enhancing innovation. There will be a likelihood that those who become creative by the works of their minds may also benefit and feel valued to even improve in future. But they have to feel protected and also need to benefit from their creative minds. This is how intellectual property comes in play as an important component.

2.4.3 Intellectual property: protecting the ideas

Intellectual property protection plays a critical role in creative thinking and ultimately in supporting creativity. For instance, the generation of ideas that results in the improved efficiency or effectiveness of a system facilitates two important aspects of creativity. There is the existence of the process. Such a process is goal-oriented; it is designed to attain a solution to a problem. Then we have the people, they are the resources that determine the solution. These individuals must be assured that the works of their minds will be rewarded and also be protected from infringement.

2.4.4 Embracing kaizen concept

The Japanese term Kaizen which refers to improvement, or change for the better, can be embedded in the creativity process. This is because it can be taken to include efforts by all employees from the top most to the lowest in cadre in striving for improvement. It strives to ensure that there is improvement of all processes and systems—including coming up with creative ideas, innovation and entrepreneurship. This is intended for the functioning of a particular organization. The concept strives for long-term superior quality and ultimately customer satisfaction [8]. In today's business world, an innovative organization can employ the Kaizen approach for evaluation of its core organizational procedures, operations and even in routine projects. Among the suggested principles of Kaizen that can address the paradigm shift include avoiding assumptions, being proactive in problem solving, creating an environment where everyone feels endowed to contribute and using creativity to find solutions to organization's needs among others.

2.4.5 Creativity and trust

Studies indicate that there is a correlation between creativity and trust. Scholars posit that as a way of nurturing creativity in order to have an innovative and supportive system in the organization, there must be provision for mistakes that might occur. In other words, there is need for crafting a safe space that provides allowance for failure. Some of the key attributes associated with firms that foster creativity and innovation are allowing individuals to acquire lessons from mistakes made. This needs some level of trust. It means the employer will have to develop patience in order to allow the employees take charge of tasks that risky enough that can lead to failure [15].

It is further argued that this may be one of the ways of getting the employees' very best ideas. For instance, it is said that those employees who are based and work in the field understand their area better than most of the other workers. What it therefore means is that by allowing them room to dip their hands in to become innovative is beneficial to them as well as to the organization.

Some firms have witnessed employee-driven creative ideas and innovative solutions that have enabled them to avoid costly mistakes thus save millions of dollars and hundreds of hours. Thus, if companies put its trust in the hands of the individuals who work hard for them daily, then the team and the firm is likely to reap benefits that far outweigh the risks and the occasional failures.

3. Conclusions the creativity gap

One scholar has opined that bridging the creativity gap may require getting the right crowd to the ideas that spark. The ideas must spur individuals to action of some sort. Involving the right crowd to get to ideas that spark may result in creativity. This is the number one skill needed to survive and thrive in the current market reality. But various organizations and individuals are yet to measure to this. Attainment of ideas which elicit excitement is not only a process, but also an Art. Many scholars and individuals are yet to develop the ability of thinking outside the box or are engaged in creating concepts or solutions. In sense, creativity is at the core of differentiating businesses, scholars and individuals. It is therefore necessary to embrace creative crowdsourcing. This could be one of the ways in which businesses can overcome their own limitations and bridge their creativity gaps by relying on the crowd's collective creative intelligence [16].

Acknowledgements

I acknowledge my own organization, the University of Eldoret for immense support in this project. The support came in during the various weeks of innovation that the university allowed the author to participate. I equally acknowledge the support of colleagues such as Eng. Antony Mbayaki for his inspiration to me in writing this article as well as my colleague Mr. Tim Koskei for his immeasurable support.

Conflict of interest

The author declares no conflict of interest.

Acronyms and abbreviations

C.E.O	Chief Executive Officer
D-Gov	Deputy Governor
KeNIA	Kenya Innovation Agency
KECOBO	Kenya Copyrights Board
KIPI	Kenya Industrial Property Institute
Mgr. (s)	Manager(s)
NACOSTI National	Commission of Science and Technology Institute
TVET	Technical and Vocational Education and Training
U.o.E	University of Eldoret
WIPO-PRV	World Intellectual Property- Swedish IP Office

Author details

Josam Nandwa Musambayi
University of Eldoret, Eldoret, Kenya

*Address all correspondence to: nandwajmusambayi@gmail.com

References

[1] Bao D. Expanding the discourse in ELT materials development through creativity and innovation. In: Creativity and Innovations in ELT Materials Development: Looking Beyond the Current Design. Multilingual Matters. Australia: Monash University; 2018. pp. 1-17

[2] Awan U, Sroufe R. Interorganisational collaboration for innovation improvement in manufacturing firms's: The mediating role of social performance. International Journal of Innovation Management. 2020;**24**(05):2050049

[3] Vermesan O, Friess P, Guillemin P, Sundmaeker H, Eisenhauer M, Moessner K, et al. Internet of things strategic research and innovation agenda. In: Internet of Things Applications-From Research and Innovation to Market Deployment. Denmark: River Publishers; 2022. pp. 7-142

[4] Auernhammer J, Leifer L, Meinel C, Roth B. A humanistic and creative philosophy of design. In: Design Thinking Research. Cham: Springer; 2022. pp. 1-15

[5] Leasa M, Batlolona JR, Talakua M. Elementary students' creative thinking skills in science in the Maluku Islands. Indonesia. Creativity Studies. 2021;**14**(1):74-89

[6] Sankaran V. Organizational leadership: Igniting creativity for sustained corporate success. In: Proceedings of the 2020 11th International Conference on Computing, Communication and Networking Technologies, ICCCNT 2020, Kharagpur, India. 1-3 July 2020

[7] Juliana NO, Hui HJ, Clement M, Solomon EN, Elvis OK. The impact of creativity and innovation on entrepreneurship development: Evidence from Nigeria. Open Journal of Business and Management. 2021;**9**(4):1743-1770

[8] Gafour OW, Gafour WA. Creative thinking skills–a review article. Journal of Education and E-Learning. 2020;**4**:44-58

[9] Rampa R, Agogué M. Developing radical innovation capabilities: Exploring the effects of training employees for creativity and innovation. Creativity and Innovation Management. 2021;**30**(1):211-227

[10] Brissaud D, Sakao T, Riel A, Erkoyuncu JA. Designing value-driven solutions: The evolution of industrial product-service systems. CIRP Annals. 2022;**71**(2):553-575

[11] Hofmann F, Jaeger-Erben M. Organizational transition management of circular business model innovations. Business Strategy and the Environment. 2020;**29**(6):2770-2788

[12] Caniëls MC, Rietzschel EF. Organizing creativity: Creativity and innovation under constraints. Creativity and Innovation Management. 2015;**24**(2):184-196

[13] Corazza GE, Agnoli S. The DA VINCI model for the creative thinking process. In: Homo Creativus. Cham: Springer; 2022. pp. 49-67

[14] Beda Z, Smith SM, Orr J. Creativity on demand–hacking into creative problem solving. NeuroImage. 2020;**216**:116867

[15] Ndemo B. The paradigm shift: Disruption, creativity, and innovation in Kenya. Digital Kenya: An entrepreneurial revolution in the making. 2017;**9**(4):1-12

[16] Laverty M, Littel C. Creativity, Innovation and Invention. In: Entrepreneurship. Houston: OpenStax; 2020. p. 15

Chapter 5

A Model of Technological Imagination and Creativity: Cognitive Task Analysis

Ruey-Yun Horng, Ching-Wen Wang, Yun-Chieh Yen and Ting-Yu Wu

Abstract

An integrated model of cognitive tasks involved in the process of a technological innovation was proposed based on these theories: 1. CDIO theory of technological innovation, 2. Wallas's creative thinking processes, 3. Khalr & Simon's theory of scientific discovery, and 4. the conceptual combination theory of imagination. The central theme of this model is the proposition that three cognitive conditions are necessary for technological imagination and innovation: 1. cross-domain knowledge, 2. simple heuristics, and 3. pattern recognition ability. Although the required domain knowledge and implementation methods are different across domains, heuristics that lead to a breakthrough at each phase of CDIO in a technological innovation are similar, with conceptual combination as the cognitive engine for generating original and imaginative ideas.

Keywords: technological imagination, creativity, technological innovation, conceptual combination, problem-solving processes

1. Introduction

From the problem conception in terms of a functional demand to finding a solution, the designing and making of the product, its manufacturing and marketing, to taking social responsibility, technological innovation is a complex problem-solving process, requiring the cooperation of a chain of different industries and experts. For example, the semiconductor industry was formed around 1960 and consists of an aggregate of companies engaged in the design and fabrication of semiconductors and semiconductor devices, such as transistors and integrated circuits (IC). It is in turn the driving force behind the wider electronics industry for such products as power electronics, consumer electronics, and e-commerce. The semiconductor market segments include networks and communications devices. A need for technological innovation in the IC manufacturing industry may be driven by demands or orders from the users, the IC design, the IC fabrication and manufacturing, IC packaging and testing, logistical supports to deliver products to customers, and user services. The size of silicon wafers determines how many IC can be produced per wafer in

IC manufacturing. The semiconductor industry has been continuously striving to increase the wafer size. Although foundries used to produce 3-inch wafers, today's common wafer size is 12 inches. Increasing wafer size is not a trivial process. In fact, silicon wafer manufacturing technologies have to be re-designed and re-engineered in order to increase the wafer size. Innovation is a common daily demand of all companies involved [1].

The CDIO theory proposed by Crawley and colleagues [2, 3] for improving engineering education pinpoints 4 critical phases in technological innovation, whether it is a new product, procedure, or system [4]:

1.1 Conceptualization

The need for a certain functional requirement of a product/system that is meaningful and valuable to the society is sensed, such as how to reduce CO_2 emissions from the production process. This raises a design problem, the next phase of innovation.

1.2 Design

The initial functional requirement is abstract in nature. It must be developed into the tangible possible designs of a physical device or system. Based on the design of the current product, the industrial designer may sketch several different designs of how the product/system will look like when new functional requirements are added. The most promising design is chosen, given current technological constraints, and a prototype is constructed.

1.3 Implementation

A prototype is merely a conceptual design presented in 3D. How the prototype of a conceived new product or system can be physically realized must be studied and implemented by a team of RD workers and engineers. That means various components of the product will be made, assembled, and contextualized into a real product that can work in a specified context.

1.4 Operation

When the first new product is created, it must be reliably reproducible with an equal quality and in a certain quantity. The manufacturing facilities must be set up so that parts can be manufactured, assembled, tested, packaged, and delivered to customers in a timely manner with a reasonable price and with supports for maintenance, repair, and even recycling.

Technological innovation is a form of complex problem-solving with humans as the primary problem solvers. Its chances of success heavily rely on deep knowledge about the product, the industry of the domain [5], and imagination and creativity. Creative ideas always spring from an individual's mind. Although CDIO may represent activities from different industries or individuals, the underlying cognitive tasks necessary for a person to come up with an innovative idea are similar. Understanding how the human mind stores, retrieves, transfers, combines, and transforms ideas in the human cognitive system is important to facilitate the generation of original and satisfactory solutions at each CDIO phase.

The purpose of this study is to propose a model of creative processes in each CDIO phase of technological innovation that incorporates Wallas's four stages of creative processes [6]. Conceptual combination theory of imagination was used to account for what happened during the incubation stage that later brought about the breakthrough ideas at the illumination stage [7]. This model of imagination and creativity in technological innovation will later be shown to be analogous to Klahr and Simon's [8] theory on the cognitive process in scientific discovery in which domain knowledge, simple heuristics, and pattern recognition are three cognitive tasks indispensable for making scientific breakthroughs.

Specifically, we propose that in order to achieve something creative, Wallas's four stages of creative problem-solving processes (i.e., preparation, incubation, illumination, and verification) will occur during each CDIO phase. The conceptual combination theory of imagination is a simple heuristic to generate imaginative and innovative ideas during the incubation stage, which then dawn on the individuals at the illumination stage. As shown in **Table 1**, this model of technological imagination and creativity can provide a handy framework for analyzing the cognitive tasks required

Phase[1]	**C**	**D**	**I**	**O**
Stage[2]/ goal	Problem finding	Solution finding	Making the first product: contextualization and specification	Building a production system
Prp.	Memory search and knowledge acquisition about problems related to: The current products and user experiences.	Memory search and knowledge acquisition about: solutions of the problem, and user experiences.	Memory search and knowledge acquisition about: methods, tools and materials for making the product, contexts of use, and user characteristics.	Memory search and knowledge acquisition about: Production methods, materials, machinery, human resources, users, markets, law, management, and etc.
Inc.	Automatic spreading activation of memory and conceptual combination.	Automatic spreading activation of memory and conceptual combination.	Automatic spreading activation of memory and conceptual combination.	Automatic spreading activation of memory and conceptual combination.
Ill.	Pattern recognition	Pattern recognition	Pattern recognition	Pattern recognition
Vrf.	Testing by logical analysis, experiment and data analysis.	Testing by building a prototype and discerning user satisfaction.	Testing by pilot runs of the product and learning levels of user satisfaction.	Testing by users of the production system and market.

[1]*C, conceptualization, D, design, I, implementation, O, operation.*
[2]*Prp., preparation, In., incubation, Il., illumination, Vrf., verification.*

Table 1.
A model of creative processes in technological innovation.

in every stage at a given CDIO phase and also for designing educational programs and procedures to facilitate engineering imagination and creativity.

2. Creative problem-solving processes in technological innovation

Creativity is a criterion for evaluating all technological innovation. Imagination is often treated as a synonym for creativity. In the present study, the distinction between imagination and creativity will be stressed so that conditions that enhance engineering imagination and creativity can be better understood. Creativity is defined as the process by which a socially valuable new product is produced [9], whereas imagination is defined as the process by which a new idea emerges in one's mind. This distinction is important because the creation of an appropriate but novel product valued by society is often a very long and complex process (e.g., the designing and manufacturing of an electrical vehicle).

Factors affecting technological creativity include not only an individual's personal characteristics such as ability, personality, and motivation but also the social, intellectual, and technological resources necessary to support such endeavors [10]. Thus, it is argued that highly creative achievement is determined by a multiplicative rule composed of many personal, social, and environmental factors. The absence of one required factor would render the creative effort futile [11]. Imagination, on the other hand, occurs in one's head and is relatively easier to manage and foster through education and training.

Klahr and Simon [8] reviewed studies on historical accounts of scientific discoveries, psychological experiments with nonscientists working on tasks related to scientific discoveries, direct observation of ongoing scientific laboratories, and computational modeling of scientific discovery processes. They proposed that cognitive processes in scientific discovery consist of three components: search through domain knowledge, use of some simple heuristics to consider and generate ideas, and pattern recognition when a solution emerges in the mind. In the present study, it is proposed that although very different kinds of knowledge are required for creative problem-solving activities during each CDIO phase, the creative-thinking processes are nevertheless similar. The conceptual combinations are the simple heuristics or cognitive mechanisms that humans use to imagine and go beyond what is given in our memory.

In brief, every episode of creative accomplishment will go through Wallas's four stages of the creative problem-solving process [6]: first is the preparation stage, during which an individual senses the problem and searches for available solutions but fails; second, in the incubation stage, because of the failure in previous problem-solving attempts, the individual puts aside the problem and shifts attention to some unrelated activities and appears to be disengaged from the problem; third, in the illumination stage, after a short or long period of incubation during which the person does nothing deliberately, the solution to the problem may suddenly pop into the individual's mind; and the last stage is verification, during which the individual performs required tests to see if the idea will indeed work.

These four stages require different cognitive engagement and social participation. For example, extensive learning and searching for information and examples of past and currently available products or practices are the major cognitive activities in the preparation stage. Implementation of the ideas and testing how they accomplish the goals usually require collaborative work and investment of material and financial support from society. Of these four stages, the incubation stage and the illumination stage occur only in an individual's head.

How the human mind worked used to be regarded as a black box. However, with recent advances in neuroscience and neuroimaging technology, the mind is no longer as black as it used to be. The present study proposes that the way the brain works during incubation to bring about illumination is through the act of imagination. Imagination is a kind of mental activity through which existing knowledge in one's associative brain networks can be activated, combined, and interpreted so that new ideas may be generated. Conceptual combination is thus proposed to be the cognitive mechanism, a simple heuristic, by which the human brain creates new ideas during the incubation stage [12].

3. Imagination: conceptual combination and interpretation

Conceptual combination refers to joining two or more concepts to generate a new idea [13]. It is believed to be the most wonderful merit of the human mind [14]. At the neurological level, conceptual combination is made possible because any neuron (concept) in human brains can be linked to many other neurons unconsciously or consciously. The meaning of an object or event is thus determined by its activation pattern, a created network, of neurons [15, 16].

In engineering, many innovative ideas were indeed conceived and created through conceptual combination. For example, the electrical bicycle is made possible by joining an electricity-powered mechanism and the manual bicycle. Personal computers originated from the idea that the traditional mainframe computer could be carried with you wherever you go, like a briefcase. The Blackberry phone was first conceived by joining of two concepts: cell phone with email [17]; mirrors were added to elevators as a means to distract people's attention from the slow speed of elevators [18]. In other areas such as language, new words and phrases are constantly created via a combination of already existing words [19, 20]. In reading comprehension, the meaning of a sentence is achieved by activation and the integration of different word meanings in a sentence so that a gist in the form of a proposition for the sentence, a macroproposition for the paragraph, and a situation model for the entire text could be derived [21, 22].

Chance or luck has always been considered an important element in creative achievement [23]. For example, Campbell's theory of blind variation, selection, and retention [24, 25] argues that random variation is the basic mechanism that organisms use to cope with an uncertain environment. For example, mutation is the major mechanism by which a virus adapts to harmful environmental factors. Humans are blessed with a brain that can perceive the environment, store learning and experiences, and retrieve the information from memory to help understand and solve new problems that arise in the present and future. By manipulating information in the brain, simulating possible environmental changes, or just letting the brain wander randomly, humans, unlike lower organisms, can think and plan their coping responses in advance. The way human brains work to create new knowledge is by a random combination of ideas in the brain. Ideas may automatically collide with each other by chance, and some patterns may emerge unexpectedly to shed light on the puzzled mind [24]. However, mere activation and association of previously stored information in the memory is not enough to generate original ideas from conceptual combination. Some novel ideas that are remote associates from other knowledge categories must be found to help one get out of a rut [26], and interpretations must be made to meaningfully link the concepts [27], either automatically or deliberately.

After reviewing extensive empirical evidence from psychological literature, Simonton [28] used the theory of constrained stochastic process to account for scientific creativity. He pointed out that the road to scientific discovery was filled with uncertainty and luck. Logics and systematic thinking do not provide much help. They are primarily useful after the discovery to provide proofs and explanations for the validity of the discovery. Based on his analysis, Simonton pointed out two factors that contributed to scientific discovery. The first is the domain knowledge of which each scientist can only sample a small portion as the target of the study. The second factor is the scientists who evaluate the existing knowledge in the field of their study and who then try to find new and useful knowledge via the combination of knowledge shared in their scientific community.

In the theory of constrained stochastic process, the random combination of ideas that leads to scientific discovery proceeds in two steps. One is ideation, which refers to the generation of possible ideas; the other is elaboration, which refers to further explanation and refinement of the ideas. Because scientists usually work on more than one project at a time, the cross-talk from these diverse projects may provide chances for novel reformulations and discoveries. Blind variations and combinations of existing knowledge are thus the brain's mechanisms to generate original ideas for different types of creativity [25].

Technological innovation is founded on scientific knowledge. The difference between technological innovation and scientific innovation is that the former is less constrained [28] because there usually exists numerous possible solutions to a technological problem. Empirical evidence shows that random variation and conceptual combination are crucial for the emergence of original technological solutions. For example, Zeng et al. [18] found that the remote association of concepts from different domains spawned more creative mash-up web service design ideas in information technology than were generated from intradomain mash-up web service design. It is thus proposed that conceptual combination is the simple heuristics to promote imagination and creativity in all four CDIO phases in technological innovation.

4. Interpretation: making sense of novel experiences

Emergence of original ideas from conceptual combination is not merely a process of memory retrieval from free association [29], neither is it the remote association [26] of concepts in one's knowledge network. When two seemingly unrelated ideas are retrieved from memory, an interpretation process is automatically activated in the human cognitive system, aiming to make sense and find meaning for the co-occurrence of these objects or events [19]. Finding a way to meaningfully connect two terms is the simplest form of relational thinking, a kind of abstract thinking [30]. It is this interpretation process that generates and adds new nodes to the memory network and makes the knowledge network grow by itself. On the one hand, because there are many possible ways to interpret the co-occurrence of concepts, objects, or events, the interpretative processes involve divergent thinking, thus making evaluation of the fluency, flexibility, originality, and elaboration of emerged ideas possible. On the other hand, the act of interpretation also requires convergent thinking because a new idea must be generated and chosen to account for the reason why these concepts co-occurred [7]. Three criteria were raised to evaluate whether the emerged new concepts from conceptual combination indeed made sense: diagnostic feasibility, plausibility, and informativeness. That is, the new concepts must be derived from distinctive feature or meaning of the given concepts, be a plausible explanation of the co-occurrence of these given concepts, and add something new when combined [31]. For example, "that lawyer is a shark" would automatically generate an image of a (greedy and reckless) lawyer with sharp teeth and biting, rather than a lawyer with a fish tail.

People use three kinds of interpretation to associate two unrelated words: 1. conjunctive interpretation, 2. property transfer interpretation, and 3. relational interpretation [27]. In conjunctive interpretation, new ideas emerge from finding property overlap of concepts to be combined. For example, "vitamin C" is the concept that emerges from the combination of the concepts "*banana–apple*." In property transfer interpretation, new concepts emerge by giving the unique property of one concept to another concept. For example, the combination of the two concepts "*lily–light*" may produce the response "lily-shaped desk lamp." In the relational interpretation, a mediating concept is introduced to link two concepts such that the initially unrelated concepts become related. For example, "*lavender–beer*" may produce a response such as "beer party in a lavender garden." Two subtypes of conjunctive interpretation were also observed [32]. Other than finding a common property for the two concepts in conceptual combination, the mapping/conjunction between two concepts can also occur at a more global level or at a structural level. This type of mapping is usually accomplished by analogy. For example, the concept of an airplane is structurally analogous to a bird generated from "*fly like a bird*." Another unique type of conjunctive interpretation is "negation;" the new concept is obtained by a negative interpretation of the common property found between two concepts, for example, "darkness" as a response counter to the property "brightness" derived from *light-rationality* pair. Originality of new concepts emerging from different types of interpretation may be quite different in terms of whether the original concepts are modified or whether the new concept entails using the original concepts to create a broader meaning.

5. Conditions that foster originality in conceptual combination

However, not all ideas that emerge from a combination of novel concepts are original [25]. Several factors might affect the outcome of a conceptual combination, including novelty or semantic distance of concepts [33, 34], the abstractness of the concepts to be combined [35, 36], types of interpretation required [37–39], age [40], number of iterations [41], or even the ontological category of the concepts. For example, Bock and Clinton [39] found that noun-noun pairs of the natural kind elicited significantly more property-related interpretations (e.g., moon-orange → round) than noun-noun pairs of the artifact kind. Likewise, noun-noun pairs of the artifact kind elicited significantly more relational interpretations (e.g., knife-bike → cutting the edge) than noun-noun pairs of the natural kind. Last but not least, recognition and decisions regarding which ideas to select and retain from a pool of interpretations are also important [8, 24].

Imaginativeness of an emerged idea is thus a function of availability and distance among the ideas to be combined in one's brain, the type of interpretation chosen in conceptual combination, and pattern recognition to identify the desired final idea. This model of technological imagination and innovation can therefore be described by integrating Klahr and Simon's [8] model of scientific discovery and Simonton's [25, 28, 42] blind variation, selection, and retention (BVSR) theory of creativity. Namely, cognitive processes involved in technological innovation include three

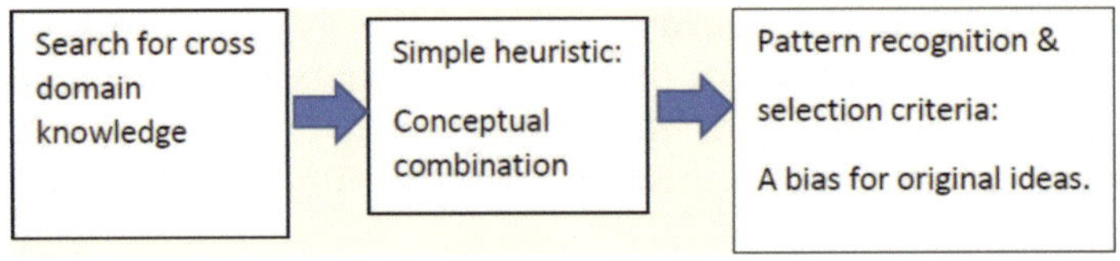

Figure 1.
Cognitive process in technological innovation.

components: 1. availability of cross-domain knowledge, where concepts from two or more than two domains need to have a chance to meet in one's head; 2. a simple heuristic, namely conceptual combination, is applied to make one or more than one interpretation to combine seemly unrelated concepts; and 3. some utility functions, or selection and retention criteria are applied so the individual can recognize and choose certain types of interpretation yielded by conceptual combination (**Figure 1**). Imaginativeness of an idea that emerges from conceptual combination is a product of these three factors within an individual. According to this model, cognitive tasks for an individual engaged in the CDIO processes of a technological innovation are analyzed as follows.

6. Cognitive task analysis of the creative processes in CDIO

Completion of CDIO tasks in a technological innovation involves the collaboration of many experts from different industries. Experts with different domain knowledge and experiences must cooperate to solve the problem at each CDIO phase and across several CDIO phases. However, for execution and completion of goals at CDIO, even though the required domain knowledge and experiences for innovation are different, the creative processes are similar. That is, at each CDIO phase, the problem is solved by going through Wallas's 4 stages of creative problem-solving processes [6]. Cognitive tasks at each problem-solving stage are summarized in **Table 1**. In the following task analysis, the conceptual combination processes are illustrated primarily with examples from the creation of the Book of Chanel N° 5 [43, 44].

The Book of Chanel N° 5 was commissioned by the owner of the perfume company Gabrielle Chanel to Irma Boom, an Amsterdam-based designer, who has made more than 250 volumes of books. About 20 percent of them are in a permanent collection at MoMA. The 300-page book has no ink. Each of the crisp white pages is embossed with a drawing or quotation that helps the story of Gabrielle Chanel unfold. The book structure is housed in a black box. The book won the Dutch Design Awards in 2013 and is part of the permanent collection at MoMA [45]. The creation of the Book of Chanel N° 5 is chosen as an example because it provides an interesting, creative example of a small but complex technological innovation that is comprehensible for most readers.

6.1 Cognitive tasks for conceptualization phase

The goal at the conceptualization phase of a technological innovation is to find a problem that is worth of future efforts. Creative ideas for a product first emerge in one's mind as a problem or a functional requirement. For example, how can I fly like a bird? Raising a question like this may drive further efforts to solve the problem.

The question must be meaningful and have potential utility to the individual. For an innovative technological problem, its meaning and utility must be recognized not only by someone, usually the users, but also by society and technological professionals, to attract resource investment. Following Wallas's theory, the cognitive tasks for emergence of creative ideas from an individual's mind at the conceptualization phase are explained as follows:

A. *Preparation stage.* For any creative and innovative task, activation of prerequisite knowledge about the product and the search for new knowledge are the primary cognitive tasks in the preparation stage. Technological innovations always build on previous technology. Features and functions are added to or removed from the previous products. For any innovative problem, there is no ready answer in the current knowledge base. One must go beyond the boundary and search in other related or unrelated domains. Cross-domain knowledge must be pooled via collaboration of the RD team. The individual's memory search can be performed either deliberately or subconsciously by the automatic spreading activation nature of the human brain [16]. A technological innovation often begins from the conceptualization of a need from users. For example, how to manufacture a wafer so that more IC could be produced per wafer? To sense the need of the problem and to solve the problem, deep knowledge and experience in wafer manufacturing is necessary. In the case of the Book of Chanel N° 5, the need for the creation of the Book of Chanel N° 5 was from the founder of the fashion company, G. Chanel, who has the knowledge of and experience with the perfume, perfume making, and the birth of Chanel N°. 5 but who is without bookmaking expertise. She turned to Irma Boom, who is an expert in bookmaking but is without expertise in the history and making of Chanel N°. 5. When Boom began working on the book, she totally immersed herself in the subject. The fashion brand provided her with as much information as possible and let it percolate until the breakthrough idea struck her. Boom spent time in Chanel's Paris apartment and studied Gabrielle Chanel's life. She made field trips to the factory and gardens and witnessed the bottling process and even joined the Chanel team as they picked roses in Grasse, a village in the Provence region of France. What she was searching was: "what is unique about Chanel N° 5?" The answer would be the key to the design idea: "what is unique about the Book of Chanel N° 5?"

B. *Incubation stage.* To come up with a new and original idea, one will certainly encounter a period of bottleneck, that is, a time when there is no idea for the solution of the problem. This period can either be relatively short or be very long, depending on the difficulty level of the problem or the knowledge state of one's memory. When one experiences failure, it is usually followed by a short or long period of breaking away. Attention is shifted to other activities. However, the brain does not stop working during this period. Remote ideas from different domains may be automatically activated unconsciously and collide with each other to form unexpected patterns or configurations of ideas [24].

C. *Illumination stage.* Once the activated memory network associated with the book's concepts (book—Chanel—perfume—fragrance—smell—invisible) was formed, as illustrated in **Figure 2**, a new idea (an invisible book) for the solution popped up suddenly after a period of incubation. Boom's conception of the unique feature of Chanel No. 5 was inspired by the nature of fragrance when

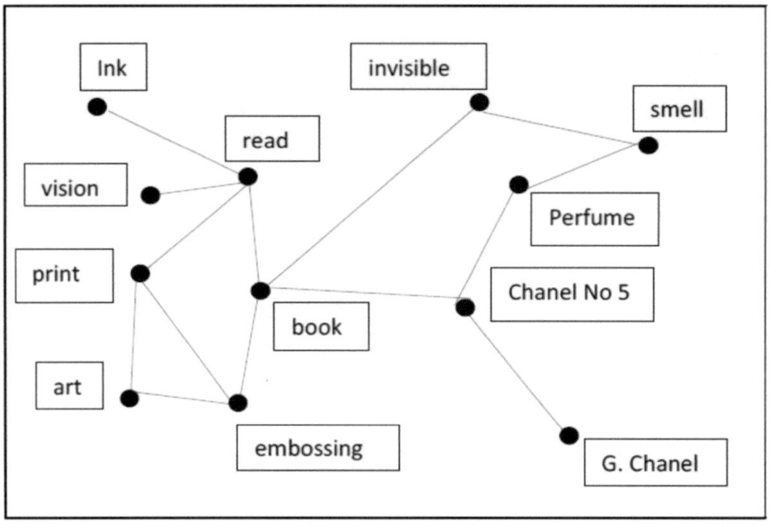

Figure 2.
Illustrated possible memory activation given book of Chanel N° 5 as cue.

Boom was in the rose garden one day and found *what she smelled there was so intense, and exciting, but not visible*. The idea for the book suddenly struck her: to create a book of Chanel N°. 5 is like to creating a book of fragrance: *a book without ink, but readable*, an analogical mapping between the book and the perfume.

D. *Verification stage.* Although the conception is abstract in nature, the idea must be subject to logical analysis and reality testing. Is it really possible to make a book without ink but for it to still be comprehensible to readers? Previous examples of books without ink such as Braille must be retrieved from memory and other external sources to verify the plausibility of the current idea.

6.2 Cognitive tasks for design phase

The abstract idea needs to be transformed into tangible, visual representations so that rough sketches of the objects in their finished form can be visualized by people, its building or construction feasibility can be examined, and alternative designs can be compared so that the better one can be chosen. The cognitive tasks involve moving from the pure abstract idea to sketches of the possible appearances of the object and then to 3-D prototypes, to envision how the new functional requirements can be materially realized. For example, how can a current 10-inch wafer manufacturing line be extended into a 12-inch wafer manufacturing line? General design knowledge and engineering knowledge about how to realize an abstract functional requirement into a concrete physical product must work together. In the case of designing a book of Chanel N° 5, the design problem is how to make a book that is invisible but readable. Boom's solution to creating a book that was without ink but readable was by *embossing*. The cognitive tasks to generate the answer "embossing" include:

A. *Preparation stage.* Knowledge of and experience with the current technology in product making is not sufficient to solve the problem; new knowledge and experience must be imported from other domains. The design is guided by the new functional requirements of the end product and existing engineering knowledge of the to-be-made product. In the example of designing the Book of Chanel N° 5, knowledge and experience in bookmaking with printing or without printing must be retrieved and compared. With a goal to create something original, previous cases of books without print must be analyzed, retained, or ruled out in the process of finding something unique. Irma Boom is an expert in bookmaking. She has created over 300 books and holds an excellent reputation for her artistic autonomy within her field. Creating a sensory tactile experience when designing and making books is very important to Boom, and she aims to inspire discovery and interaction. She analyzes every little detail in bookmaking to maximize a book's engagement potential in contrast to its digital counterparts [45].

B. *Incubation stage*. After extensive review of previous examples of books without ink, some original ideas must be generated. Many ideas and experiences regarding bookmaking will be searched and then abandoned. A period of mental block and sense of failure are inevitable. A period of breakaway from the problem is called for. However, the combination of concepts: book—no print—readable may subconsciously keep activating experiences and knowledge about books or reading under special circumstances such as for the blind, the hearing impaired, and so on. Consequently, Boom's experiences of making book covers by embossing were activated in the memory network in the incubation stage. **Figure 2** is a simplified illustration of possible concepts activated during the process of the information searching in the brain.

C. *Illumination stage.* The idea of using embossing to make the book pages emerged! It could be used to make the content of the book semi-invisible but still readable.

D. *Verification stage.* Designing a technological product is half-conceptual half-physical in nature. A prototype must be made to communicate the idea in a physical form. Boom has long used embossing on the cover of her books, but using it as the only source of printing was a new concept with unique challenges. A prototype must be produced to show how the book would appear to G. Chanel. It is a miniature of the product to be created. It shows the structure and configuration of the components of the product in its finished form. The plausibility of making the real object can then be evaluated and discussed among experts, engineers, and users. Revisions and redesign are relatively easy at this stage.

6.3 Cognitive tasks in implementation stage

A prototype of the product is still conceptual in nature. Can the prototype design work in reality? A real, physical object must be built by RD engineers. Technological constraints are the major challenge at this phase. The nature of engineering work is constraint satisfaction or contextualization [46]. For example, to create an electrical bike, adding a power system with a power control system and user interface mechanism to a bike body is the major challenge. Moreover, the final product must be contextualized to the persons and conditions under which the bike is to function. Every detail must be specified before the product can be made. Engineering knowledge, industrial knowledge, and knowledge of the potential markets and users must be searched and available for consultation. The goal of the implementation stage is to make a real object work as designed and that can operate properly in a prescribed context. The making and manufacturing of the first technological product is not only a very tedious and time-consuming process, but it is also very expensive. Other than materials and machinery for making the object, a variety of engineering knowledge, equipment, and procedures must be recruited. In the case of building the first electrical bike, technological problems involve making of the bike body, battery, power control device, man-machine interface device, and other auxiliary parts. Most important of all is putting these elements together and making sure that the bike can ride for a reasonable distance with reasonable labor and speed. RD workers and experts in the related domains must be recruited and work together to solve the new problems involved in making the first instance of the new electrical bike.

A. *Preparation stage.* In engineering, the product is first broken down into components and parts, each of which is manufactured separately. Different lines of engineers work at about the same time, and their outputs are then integrated to form the newly innovated product. New designs invariably create new technical problems for making the product. For example, embossing is commonly done with materials that are hard enough to endure the press. How can embossing be done with a sheet of thin and fragile paper? How to bind the book without damaging the content? New materials and/or new embossing techniques must be found to solve the problems. The size, weight, and content of the book may also affect how many pages of embossed content can be bound. New solutions must be found. In Boom's case, making an embossed book involved combining both printing and embossing techniques. Using embossing as the only source of printing was a unique challenge for the printing industry.

   Typically, books are bound and cut, but the pressing process would render the embossing flat, so it was necessary to figure out a different way to ensure the subtlety of Boom's designs would keep their form. The publisher ended up using an old letterpress machine, with the ink removed. Each page was first designed on an aluminum plate and turned into a mold that the pages would then press against. Numerous technical details like this one needed to be surmounted in making the first book. The solutions needed to be found before a first copy of the Book of Chanel N° 5 could be produced in its final completed form. These cognitive tasks cannot be performed by any individual alone. It requires the combined efforts of many experts from different domains. Some problems may need to be fed back to the conceptualization and design phases for reworking.

   Knowledge about the making of a product is thus distributed among different experts. Therefore, collaborations among individuals are essential. For example, more than one thousand steps are involved in wafer fabrication. It is not possible for one person to have the complete knowledge needed to create a new wafer manufacturing line. In the case of bookmaking, other than how to print the content of the book onto pages, the content of the book must be written, and the size, weight, thickness, and anticipated demands of the targeted readers and so on must be specified before the book can materially exist.

B. *Incubation stage.* A period of breaking away from the problem and the opportunity to exchange ideas among different experts are critical for breakthroughs in solution finding. Problems and a failure to find the proper solution may occur for each subproblem. The individuals who work on the problem may have to wait for some period before another's problems are solved. Communication among RD teams and their scientific and technological communities becomes vital. Various solutions are searched and activated and shared in the brains for random variations. Conceptual combination can thus occur within one's brain or between multiple brains via communication.

C. *Illumination stage*. Solutions for a given problem always emerge first from a team member's brain. The ideas must be voiced and heard and have a chance to be discussed. In the case of Chanel N° 5, the final book is 5 cm thick, a nod to the perfume's name, and each design was hand drawn. The 300-page book was printed devoid of ink. It was instead embossed with text and images, creating a semi-invisible narrative for Gabrielle Chanel. Although there was a temptation to infuse the pages with the smell itself, Boom rejected the idea because it was too literal, too obvious. For the readers, the concentration is on the images, text, and tactile sensation when one leafs through the book, creating an experience through which one can almost smell the perfume. To Boom, that is much more interesting and thought-provoking than the real smell of the perfume. The final product is a book filled with solid white, textural pages. The content of the book was taken from the world of Mlle, a book on Coco Chanel: "*Mademoiselle: Coco Chanel and the Pulse of History*" [44, 47].

D. *Verification stage*. The reliability and quality of each technological innovation must be tested via experimentation, running simulation, or field testing to make sure each part functions as designed and meets the end user's demands. In the Book of Chanel N° 5's case, the final book had to be semi-visible but completely readable by the readers and, at least, be satisfactory to G. Chanel.

6.4 Cognitive tasks in operation phase

The ultimate goal of a technological innovation is to compile and assemble a production procedure so that the product can be reproduced with reliability and flexibility and in a certain quantity upon order at any time. The possibility of manufacturing automation is also a concern. It is not just the act of creating and making a new technological product; it is the act of creating a new industry. For example, more than one thousand steps are involved in the wafer manufacturing procedure. A huge quantity of material and human resources must be invested to create a production line. Problems such as inventory control, testing, packaging, delivery, recycling, and customer service are also essential components of production line management. Cognitive tasks in the operation phase include working out every detail about the routine manufacturing, packaging, testing, and delivery of the finished product to end users while making sure the product can function properly for them. In a manner similar to the implementation stage, many sectors and people are involved in the execution of the tasks. For example, in the semiconductor industry, a wafer manufacturing line consists of an engineering chain, a supply chain, a manufacturing execution system, an equipment engineering system, and an internet system to support the interactions of these separately operated systems. Problems, such as a bottleneck in scheduling, machine failure, defects in products, or consumer complaints, need to be dealt with constantly. Daily routine operations are accomplished by many people and facilities. Some mechanical failures or quality problems may call for innovative solutions, redesigning, or reengineering. The whole cycle of CDIO may be repeated again and again. In addition, with use, users can detect problems that were missed by the designers or manufacturers. Their feedback provides invaluable opinions for product improvement and innovation. When setting up the production system for the first time, the creative processes are described as follows:

A. *Preparation stage.* The goal is to search for possible methods to manufacture the product for the market. Building a reliable, flexible, efficient, and effective manufacturing system is the goal at this phase of product innovation. User friendly, customer satisfaction, safety, ease of maintenance, and environmentally friendly are the ultimate criteria for the evaluation of the success of a product system. For example, a production system for a 5-inch wafer is not the same as the production system for a 3-inch wafer. Many new problems arise for each wafer size. They may be identified not only by experts in manufacturing and industrial management but also by operators at the work site, users, or lawmakers. Besides, because of the constantly evolving nature of science and technology, there is always a continual demand for improvements in production methods. These problems need to be solved by RD engineers or domain experts. In the case of making the Book of Chanel N° 5, numerous problems could arise regarding the fragility of an embossed book, such as how to bind the book? How to repair or replace a damaged page? How to handle the book so that the embossed content will not be accidentally erased? How long would the embossed content stay legible? How to store or ship the book so that its contents will not be damaged? These problems may come as feedback to the experts in the conceptualization, design, or implementation phases for clarification and solutions. Or they might be solved by personnel such as production engineers, quality control engineers, or industrial engineers at the work site. The search for knowledge regarding production methods, equipment, materials, and procedures is again the first step to take here.

B. *Incubation stage.* A period of rest and separation from the problems at hand is necessary for remote associations in the memory to be activated and reach the activation level. Engineers and technical support staff need to come up with plans for a manufacturing plant. When actually building the production line, new technical problems may be detected and sent to RD engineers for solutions. Some problems may need to be sent to implementation, design, or even conceptualization phases for further clarification and ideation. In essence, the knowledge for problem-solving is again distributed among many different experts. Time and opportunity for the exchange of ideas and experiences are critical for remote conceptual combinations to occur.

C. *Illumination stage.* Once the remote idea reaches its activation level, it may pop out of an individual's head after some period of incubation. The idea must have a chance to be heard, discussed, and accepted by the team. For example, to protect the embossed content of the Book of Chanel N° 5, the book structure is housed in a black box so that the content of the book will not be damaged during handling and delivery.

D. *Verification stage*. The efficiency, effectiveness, safety, and environmental sustainability of the manufacturing procedure and the quality of its outputs must be constantly evaluated. User satisfaction, the success of sales, and employee satisfaction and safety are the final tests of the success of technological innovation.

Other than the test of the market, the Book of Chanel N° 5 is now in a permanent collection at the Museum of Modern Art in New York City.

As summarized in **Table 1**, the above cognitive task analyses show that CDIO represents different domains of knowledge involved in the innovation process of a new technological product. From the conceptualization of a need for a new product, to the design of the physical product, to making the first instance of the product, and, finally, to creating a manufacturing line for reproducing the product, the diversity and complexity of the tasks involved increase tremendously with the progression of the innovation. Cross-domain knowledge and collaboration among experts and professionals are crucial for the success of technological innovation. The criteria and methods to validate the viability of the innovation outcome for each phase of CDIO are different. Heavy reliance on cross-domain knowledge, user experiences, and cross-domain collaboration are unique features of technological innovation. Nevertheless, the creative cognitive processes, namely, the incubation stage and illumination, are similar across the 4 phases and are also similar to other types of creative works such as scientific discovery or art. Specifically, creative ideas are always conceived within and emerge from an individual's mind. However, it is very likely that the remote associates for conceptual combination may come from another person's mind. The opportunity to communicate and share thoughts with one another is important in technological innovation. A period of a brief or long pause from the current domain area is necessary for the mind to shift its focus of attention and wander away from the current problem so that remote associates from another domain may gain access to the conscious state. Thus, room to endure the ambiguity and uncertainty of the situation must be provided. An illumination of a solution popping out of the mind occurs when one of the team members recognizes the potential of relating the current problem to a remote associate, and an interpretation is formed and acknowledged by the team. This creative cognitive process can be succinctly delineated by a slight modification of Klahr and Simon's [8] model of scientific discovery, shown in **Figure 2**.

7. Conclusions and discussion

In sum, the model shows three conditions for the generation of imaginative and creative technological innovation: 1. the importance of cross-domain knowledge is stressed, not just because an innovative solution, by its definition, cannot be found in the original domain of the product but also because the nature of technological innovation, more often than not, involves combinations or incorporation of new functions from other domains. For example, a smartphone is a combination of an original mobile phone and internet functions. 2. conceptual combination is viewed as a primitive heuristic for information searching and interpretation. The human ability to imagine and create is rooted in the brain's capability to randomly combine information stored in memory and make sense of these novel conceptual combinations. It highlights the importance of not only memory search and retrieval but also cognitive mechanisms to make sense of the meanings implied by the remote associations of the concepts. Many heuristics and methods for innovative ideas such as Triz [48], SCAMPER [49], and morphological analysis [50] are just more elaborate and structured methods of conceptual combination. 3. Pattern recognition is a cognitive task not very well explained before in terms of how an original idea can be recognized and selected as a plausible solution to the problem. Here, it is proposed that the ability to recognize and select the outcome from conceptual combinations is a joint function of the goal, the interpretation efforts, and the value (bias) held by the individual and the team members. Recognition of a solution in its primitive form is shaped by the goal one is searching for in the creative process. Without it, one can be totally blind to the opportunity [51]. However, the acceptance and adoption of technological innovation are also determined by the group's value about what the most desirable outcome is and how much risk (the investment) the group would take for its success in the making and selling of the product. Cultural and social factors may also come into play at the verification stage of every phase of CDIO [25, 42, 52].

The plausibility of the model awaits empirical verification. It cannot be tested directly. However, its theoretical and practical implications can be tested by logical analyses, experiments, field studies, or case studies. Because technological innovation is an application of scientific knowledge to the development of artifacts for human use, the importance of scientific knowledge and user experience for innovation in every phase of CDIO is beyond doubt. The nature of technological innovation itself is a combination of different technological products employed to produce a function that has higher utility than the previous one. Conceptual combination is the most primitive form of creative thinking. Empirical studies that examined the effects of conceptual combination in technological innovation were reported in the earlier sections of this paper. Studies can also be done to examine the effects of diversity of a team's knowledge and its team members' imaginative abilities on product innovation. For example, in Wang, Lu and Li's study [53], data were drawn from 49 dyads who were the finalists out of 120 teams of a collegiate saw-design competition. Their task was to conceive and design an unusual use of the saw. The ideas behind the sketches of their design were presented and scored by 3 professors and 2 design professionals according to three criteria: inventiveness (60%), clarity of conceptualization and presentation (20%), and creative strategy for competition and marketing (20%). Results showed that participants' imagination score measured by a conceptual combination test, efficiency, effectiveness of communication between the dyad, and heterogeneity of the team composition all contributed positively to design performance. The interaction between the imagination score and the heterogeneity of the team suggested that the dyads with higher imagination scores produced more creative designs when their collaborators were from a more different domain. In addition, a behavioral measure of imagination was constructed based on conceptual combination theory with acceptable reliability and validity [7]. The test scores were found to be able to predict design students' design performance more than the originality measure of divergent thinking ability. Methods of training to enhance engineering imagination based on conceptual combination theory have also been designed and can be incorporated into engineering education [32].

One practical implication that deserves special attention for technological education and the industry is that chance plays a role in creative processes, but it does not come without cost. The magic of incubation and illumination suggests that, given sufficient motivation and prerequisite conditions, human minds may continue to freely search and combine ideas even subconsciously. In an industry, the absorptive ability of a firm, most likely an effect spilled over from its leaders, affects the innovativeness of the firm [54]. In engineering education, efforts should not be limited to only acquiring the CDIO knowledge necessary for technological innovation but also include recognition and cultivation of the important environmental conditions for innovative ideas to be brooded upon, pop out, and be recognized and selected.

Conceptual combination is a simple heuristic for generating imaginative ideas. It is made possible by the way human neurons may randomly combine with other neurons and generate new links between existing nodes. This automatic bottom-up process is accompanied by a top-down interpretation process that makes sense of the possible relation(s) between or among originally unrelated nodes (concepts) and generates new concepts. Life experiences and knowledge are indispensable components in the process, with the goal as the engine that ignites all the processes, decisions, and steps taken in CDIO phases of a technological innovation. A technological innovation project involves problems to be creatively solved for the first time in history. The final product is the joint effort of many experts and professionals. For a firm, heavy workload and time pressures are killers of a firm's opportunity to think freely and the tendency to interact with one's coworkers in good humor. In engineering education, providing training regarding the complete CDIO process in technological innovation and execution enables future engineers to gain a broader perspective of their tasks at hand and prepares them with a mind that can solve problems more creatively. The bottleneck in the use of conceptual combination lies firstly in the time-consuming process of knowledge acquisition, not only within one's own application domain but also across other domains, and secondly in the ability to come up with original interpretations, a kind of abstract thinking and hypothesis generation, for a novel conceptual combination. Broad interests, curiosity, and a quest for knowledge are few of the important traits to be cultivated in engineering education in order to achieve minds that value novel, creative ideas and that are willing to play with ideas imaginatively, take reasonable risks, and use labor wisely to implement new ideas.

Acknowledgements

This work was in part supported by Taiwan's Ministry of Science and Technology under Grant Number: MOST 107-2511-S-009-00-MY3.

Author details

Ruey-Yun Horng[1*], Ching-Wen Wang[2], Yun-Chieh Yen[2] and Ting-Yu Wu[3]

1 National Young-Ming Chiao Tung University, Hsinchu, Taiwan

2 National Chung Hsing University, Taichung, Taiwan

3 Minghsin University of Science and Technology, Hsinchu, Taiwan

*Address all correspondence to: ryhorng@nycu.edu.tw

References

[1] Xiao H. Introduction to Semiconductor Manufacturing Technology. 2nd ed. Englewood Cliffs, NJ: Prentice Hall; 2012. 698 p

[2] Berggren K-F, Brodeur D, Crawley E, Ingemarsson I, Litant WTG, Malmqvist J, et al. CDIO: An international initiative for reforming engineering education. World Transactions on Engineering and Technology Education. 2003;**2**:49-52

[3] Crawley E, Malmqvist J, Ostlund S, Brodeur D, Edstrom K. Rethinking Engineering Education: The CDIO Approach. New York: Springer; 2014. 331 p

[4] Smulders A, Kamp A, Fortin C. The CDIO framework and new perspectives on technological innovation. In: Proceedings of the 14th International CDIO Conference, Kanazawa Institute of Technology; June 28-July 2, 2018; Kanazawa, Japan. Available from: http://ds.libol.fpt.edu.vn/handle/123456789/2447

[5] Newell A, Simon H. Human Problem Solver. New York: Cambridge University Press; 2019. 938 p

[6] Wallas G. The Art of Thought. New York: Harcourt, Brace; 1926. 314 p

[7] Horng R-Y, Wang C-W, Yen Y-C, Lu C-Y, Li C-T. A behavioral measure of imagination based on conceptual combination theory. Creativity Research Journal. 2021;**33**:376-387

[8] Klahr D, Simon H. Studies of scientific discovery: Complementary approaches and convergent findings. Psychological Bulletin. 1999;**125**:524-543

[9] Sternberg RJ, editor. Handbook of Creativity. New York: Cambridge University Press; 1999. 490 p

[10] West MA. Sparkling fountains or stagnant ponds: An integrative model of creativity and innovation implementation in work groups. Applied Psychology: An International Review. 2002;**51**:355-387

[11] Simonton DK. Talents and its development: An emergenic and epigenetic model. Psychological Review. 1999;**106**:435-457

[12] Fordor JA, Pylyshyn S. Connectionism and cognitive architecture: A critical analysis. Cognition. 1988;**28**:3-71

[13] Ward TB, Kolomyts Y. Creative cognition. In: Kaufman JC, Sternberg RJ, editors. The Cambridge Handbook of Creativity. New York: Cambridge University Press. p. 175-199. DOI: 10.1017/9781316979839.011

[14] Gabora L, Kitto K. Conceptual combination and the origins of complex cognition. In: Swan L, editor. Origins of Mind. Berlin: Springer; 2013. p. 361-381

[15] Benedek M, Jurisch J, Koschutnig K, Fink A, Beaty RE. Elements of creative thought: Investigating the cognitive and neural correlates of association and bi-association processes. NeuroImage. 2020;**210**:116586. DOI: 10.1007/978-981-10-0983-9_33

[16] Collins AM, Loftus EF. A spreading-activation theory of semantic processing. Psychological Review. 1975;**82**:407-428

[17] Thagard P, Steward TC. The AHA! Experience: Creativity through emergent bonding in neural networks. Cognitive Science. 2011;**35**:1-33

[18] Zeng L, Proctor RW, Salvendy G. Fostering creativity in product and

service development: Validation in the domain of information technology. Human Factors. 2011;**53**:245-270

[19] Middleton EL, Rawson KA, Wisniewski EJ. How do we process novel conceptual combinations in context? The Quarterly Journal of Experimental Psychology. 2011;**64**:807-822

[20] Maquire P, Wisniewski EJ, Storms GA. Corpus analysis of conceptual combination. In: Vosniadou S, Kayser D, Protopapas A, editors. Proceedings of European Cognitive Science Conference 2007. New York: Taylor & Francis; 2017. p. 417-425

[21] Kintsch W. Comprehension: A Paradigm for Cognition. New York: Cambridge University Press; 1998. 480 p

[22] Kenesei I. The role of creativity in the cognitive turn in linguistics in aging. International Review of Pragmatics. 2013;**5**:271-292

[23] Austin JH. Chance, Chase and Creativity: The Lucky Art of Novelty. Cambridge, MA: The MIT Press; 2003. 245 p

[24] Campbell DT. Blind variation and selective retentions in creative thought as in other knowledge processes. Psychological Review. 1960;**67**:380-400

[25] Simonton DK. Scientific creativity: Discovery and invention as combinatorial. Frontiers in Psychology. 2021;**12**:721104

[26] Mednick SA. The associative basis of the creative process. Psychological Review. 1962;**69**:220-232

[27] Wisniewski EJ. Construal and similarity in conceptual combination. Journal of Memory & Language. 1996;**35**:434-453

[28] Simonton DK. Scientific creativity as constrained stochastic behavior: The integration of product, person, and process perspectives. Psychological Bulletin. 2003;**129**:475-494

[29] Osborn AF. Applied Imagination: Principles and Procedures of Creative Problem Solving. New York: Scribner; 1953. 317 p

[30] Gentner D. Bootstrapping the mind: Analogical processes and symbol systems. Cognitive Science. 2010;**34**:752-775

[31] Costello FJ, Keane MT. Efficient creativity: Constraint-guided conceptual combination. Cognitive Science. 2000;**24**:299-349

[32] Horng R-Y, Wang C-W, Li C-T. An engineering imagination training paradigm based on conceptual combination theory. In: Proceedings of the 42th SEFI Annual Conference; 15-19 September 2014; Birmingham, UK. p. 0026. 8 p

[33] Wan WWN, Chiu C-Y. Effects of novel conceptual combination on creativity. Journal of Creative Behavior. 2012;**36**:227-240

[34] Kao CY. How combining opposite, near-opposite, and irrelevant concepts influence creativity performance. Psychology of Aesthetics, Creativity, and the Arts. 2019;**13**:24-35

[35] Gilson LL, Madjar N. Radical and incremental creativity: Antecedents and processes. Psychology of Aesthetics, Creativity, and Art. 2011;**5**:21-28

[36] Ward TB, Patterson MJ, Sifonis CM. The role of specificity and abstraction in creative idea generation. Creativity Research Journal. 2004;**16**:1-9

[37] Scott GM, Longergan DC, Mumford MD. Conceptual combination: Alternative knowledge structure, alternative heuristics. Creativity Research Journal. 2005;**17**:79-98

[38] Kohn NW, Paulus PB, Korde RM. Conceptual combinations and subsequent creativity. Creativity Research Journal. 2011;**23**:203-210

[39] Bock JS, Clifton C Jr. The role of salience in conceptual combination. Memory & Cognition. 2000;**28**:1378-1386

[40] Marshal N, Coblentz S. Creative interpretations of novel conceptual combinations in aging. Creativity Research Journal. 2014;**26**:158-164

[41] Chan J, Schunn CD. The importance of iteration in creative conceptual combination. Cognition. 2015;**145**:104-115. DOI: 10.1016/j.cognition.2015.08.008

[42] Simonton DK. The blind-variation and selective-retention theory of creativity: Recent developments and current status of BVSR. Creativity Research Journal. 14 Apr 2022. 20 p. DOI: 10.1080/10400419.2022.2059919 [In press]

[43] Available from: https://www.youtube.com/watch?v=0yyPbnYx-IA&ab_channel=PrintPowerEurope

[44] Available from: https://go.distance.ncsu.edu/gd203/?p=68600

[45] Available from: https://en.wikipedia.org/wiki/Irma_Boom

[46] Altshuller GS. Creativity as an Exact Science: The Theory of the Solution of Inventive Problems. London: CRC Press; 1984. 320 p

[47] Garelick RK. Mademoiselle: Coco Chanel and the Pulse of History. New York: Random House Trade; 2015. 624 p

[48] Ilevbare IM, Probert D, Phaal R. A review of TRIZ, and its benefits and challenges. Technovation. 2013;**33**:30-37

[49] Serrat O. The SCAMPER technique. In: Serrat O, editor. Knowledge Solutions: Tools, Methods, and Approaches to Drive Organizational Performance. New York: SpringerOpen; 2017. p. 311-314

[50] Sunder V, Ganesh LS, Marathe RR. A morphological analysis of research literature on Lean Six Sigma for services. International Journal of Operations & Production Management. 2019;**38**:149-182. DOI: 10.1108/IJOPM-05-2016-0273

[51] Most SB, Scholl BJ, Cliford ER, Simon DJ. What you see is what you set: Sustained inattentional blindness and the capture of awareness. Psychological Review. 2005;**112**:217-242

[52] Amabile TM. Creativity in Context: Update to the Social Psychology of Creativity. New York: Routledge; 2018. 336 p

[53] Wang C-W, Lu C-Y, Li C-T. The effects of imagination and team communication on design performance. International Journal of Technology and Engineering Education. 2015;**12**:1-13

[54] Wang Y-L, Wang Y-D, Horng R-Y. Learning and innovation in small and medium enterprises. Industrial Management & Data System. 2010;**110**:175-192

Chapter 6

Creativity, Innovation and Shariah Compliance by the Malay Middle Class Women Entrepreneurs in Muslimah Fashion Industry in Selangor, Malaysia

Mashitah Sulaiman and Mohd Nazir Ahmad

Abstract

The element of creativity is a vital and interrelated subject to innovation of a product especially in the context of entrepreneurship and business. It is also mentioned by many Muslim and Western scholars in the development of an excellent civilization. Thus, this chapter discusses the importance of creativity and innovation in the pursuit of Muslimah fashion among the Malay Muslim women entrepreneurs of the middle class in Selangor. This chapter also examines the factors that led to the involvement of women entrepreneurs in the Muslim fashion industry. In addition, this study also analyses the element of creativity and Shariah compliance in innovating Muslimah fashion among Malay women entrepreneurs in Selangor, an urban state of Malaysia. This study employs a qualitative approach by using interview techniques or oral history through purposive sampling. The sample was selected based on the criterion sampling and a total of ten (10) informants were involved in the study. Data were analysed descriptively using content and thematic analysis. In addition, observation methods are also utilised to strengthen data collection of the study. The findings indicated the importance of creativity and innovation in the production of Muslim fashion and design products.

Keywords: creativity, innovation, Shariah, middle class, entrepreneurs, civilization

1. Introduction

The creativity and innovation are essential elements which bloom the Muslim and Hijab fashion industry which is becoming a lucrative and a productive industry worldwide, in particular, Malaysia [1]. The emergence of Muslim fashion industry has been a product or outcome of the rise of the Islamic Resurgence in Malaysia since 1970s. The recent growth of the Muslim Hijab fashion industry is fundamentally associated with the rise of Islamic awareness on the Muslim-Shariah compliance in Muslim-majority countries, not the rise of capitalism as claimed by Farhad [2]. Inside the booming of Muslim fashion industry, the global fashion industry is now recognising the credibility of Muslim designers with Islamic faith to reflect the dynamic of Muslim fashion industry.

Such phenomenon can be seen not only in the Muslim countries like Malaysia, but also in Europe. What's long been referred to as the "modest" fashion of Muslim women has grown into a major industry generating hundreds of billions of dollars annually in the United States of America (US). For the first time, U.S. museum exhibited the Islamic culture within the context of fashion and it was opened in San Francisco's de Young Museum in 2018. Jeffrey Brown [3] reports on why Muslim fashion that has previously been ignored is now gaining new attention. "It's a fashion show in a museum, with an unusual focus. This is an exhibition of contemporary Muslim fashion, the first of its kind in the U.S., put together by the de Young Museum in San Francisco".

The hijab fashion industry can be considered as the fastest growing consumer segment in the world and is viewed by designers and entrepreneurs as a critically significant segment in Malaysia [1]. Therefore, this paper aims to explore the contributing factors which lead to the involvement of the Malay Muslim women entrepreneurs in the Muslim fashion industry within the boundaries of Islamic principles. This paper also discusses the importance of creativity and innovation in the pursuit of the Muslimah fashion industry among the Malay Muslim women entrepreneurs of the middle class in Selangor. Additionally, this study also analyses the element of creativity, innovation and Shariah compliance in the Muslimah and hijab fashion trends among Malay women entrepreneurs in Selangor, an urban state of Malaysia.

2. Literature review

2.1 Relationship between creativity, innovation and entrepreneurship

Creativity, innovation and entrepreneurship are interrelated concept that could be abridged in a well-defined and clear association. Discussions on creativity and innovation are pertinent in research work because they are viewed as key components in the development of competitive advantage [4]. Creativity can be defined in many ways. Some researchers define it as a thought, and some researchers consider it as a process. Thus, creativity is defined as creating new and effective ideas. Being new refers to the purity and originality of an idea ([5], p. 74).

According to Oxford Dictionaries (2012), creativity is the use of imagination or original ideas to create something. Being creative requires constant effort on the part of the individual to refine solutions and ideas [6]. Secondly, creativity requires two basic elements which are originality and effectiveness ([7], p. 92). Originality is crucial for creativity; however, it is not sufficient. Ideas and merchandise that are basically original would possibly be useless very well. So again, originality is no longer by itself to be acceptable for creativity. Original things must be too advantageous to be creative. Like originality, effectiveness takes more than a few forms. Effectiveness may additionally take the structure of value. This label is pretty clear in the economic research on creativity; it is describes how initial and beneficial merchandise and thoughts rely on the current market, and extra clearly on the price and assets [7].

Other than that, creativity is something that happens when an individual creates or produces something that is unique as well as applicable, relevant, imaginative, innovative, or persuasive [8]. According to George and Zhou [9], creativity can be defined as the formation or creation of ideas, clarifications or solutions that are relevant and practical [10].

Creativity is the outcome of an effort when an individual creates or produces something that is unique as well as applicable, relevant, imaginative, innovative, or persuasive [8]. Creativity is also the ability to make or bring into existence something new, whether a new solution to a problem, a new method or device, or a new artistic object or form ([11], p. 721).

According to a humanistic viewpoint, the creative person has the awareness and skills necessary to respond to catastrophe in transforming ways [12, 13]. As a result, the creative person can be seen as someone who is working towards self-actualization and acquiring traits that are connected to mental health, such as subjective well-being, resilience, optimism, quality of life, and other elements stressed by positive psychology [12]. Thus, creativity also can be summarised as the tendency to produce or recognise the ideas, other options, or possible possibilities to solve the problem, communicate and entertain us and other people. For that reason, in order for something to be creative it must meet the initial criteria of novelty and then prove to be appropriate, generative or influential to reach a higher status of creativity ([8], p. 3).

The relationship between creativity, innovation and entrepreneurship has been explained by Muhamad Izzuwan Shah et al. [14], Nakano and Wechsler [12] and Goh [15]. Innovation and creativity are often regarded to as the soul and the heart of a business [4]. The capacity to generate fresh concepts and find novel approaches to issues and opportunities is referred to as creativity ([14] Porter, in [16], p. 12). Innovation, on the other hand, is the capacity to use original approaches to challenges and opportunities in order to improve people's lives or to benefit society. Innovation is the key challenges to economic growth and prosperity [17], especially in the context of dealing with the post Covid-19 pandemic [6].

Creative thinking is the cornerstone of innovation. Creativity is necessary for innovation, but it is not always enough. Innovation is the implementation of creative inspiration [4]. Innovation requires the adoption of new technology or management techniques within an organisation to accomplish a specific operational improvement [6, 18].

Many categories of innovation could be traced, such as administrative innovations, incremental products, drastic products, incremental processes, and radical processes [19, 20]. The object, sector, volume, and strength all affect the different types of innovation. They are unconstrained, autonomous, and possess distinguishable characteristics. These are: market innovation, technological innovation, organisational innovation, and product innovation [19, 21].

Although in an entrepreneurial sense there should also be a subsequent link to innovation and profitability in monetary and social terms ([22], p. 50), creativity has also been considered as the production of ideas or things that are fresh and potentially beneficial [23].

Therefore, entrepreneurship is the result of a process that applies both creativity and innovation to capitalise on marketplace opportunities in highly competitive market and global economy [15, 19]. Creating value for business and social communities by combining special public and private resources to take advantage of economic, social, or cross-cultural opportunities in a changing environment is another definition of entrepreneurship [22].

Numerous studies have shown a link between corporate creativity, innovations and survival. The success of any business depends on creativity and innovation [24]. Innovation can be regarded as the primary driver behind the survival and continuation of businesses; it fosters their development and progress and increases their chances of success in the future [25]. In summary, both creativity and innovation skills are essential in particular, to promote human potential by eliciting positive aspects of the individual [12].

2.2 The importance of creativity and innovation in entrepreneurship, business and global market

The importance and the role of creativity and innovation in entrepreneurship and business have been critically discussed by many researchers in the previous studies [5, 6, 14, 22, 26, 27]. In the recent changing global economy, the nature of business is moving from knowledge based to electronic economy or e-economy activities characterized by creativity, innovation, entrepreneurship and imagination [22, 28, 29]. This could lead to increase competition in more business opportunities resulting from the increase impact of globalisation and technology [30].

Entrepreneurs can take advantage of these opportunities by using their creativity in ways that provide their business a competitive edge. It can support innovation and commercial expansion while also having a positive social impact at large ([22, 26, 31], pp. 49–50). It is crucial to emphasise creativity as a tool for future economies since innovation plays a part in successful economies [32]. The importance of creativity and innovation in producing new and innovative business ideas benefits and adds value to companies more. Innovation and creativity are now without a doubt necessary to enable a firm last a long period and remain competitive [14]. Therefore, incorporating creativity and innovation into the agenda for entrepreneurship development is necessary for a nation to succeed and have on-going economic progress ([4], p. 1745).

In 1993, Synetics carried out some research about innovations in major business in the United States of America. Similar findings come out of research by the Department of Trade and Industry in the United Kingdom ([16], pp. 16–17). The findings indicated that the gap between what leading corporations say about innovation and what they did. The gap is large when 80 per cent of US companies said that innovation is very important to their business, but only 4 per cent say that they are good at it. These are clear indications of a consistent connection between a company's commitment to innovation and its success in the marketplace. The findings that support this contention could be categorised into sales, profits, market share and innovation budget. In order to increase a company's chances of becoming a high-performing and successful organisation, the research suggests five fundamental techniques. The first is to increase the effectiveness of meetings. Second, launch official innovation initiatives or strengthen those that already exist. Third, look for outside inspiration. Fourth, encourage better communication and teamwork. Fifth, recognise what consumers and clients will request and want in the future ([16], p. 17).

Juliana et al. [4] investigate the impact of creativity and innovation on the growth of entrepreneurship. Prior to doing so, they combine the knowledge from numerous works of literature on creativity, innovation, and entrepreneurship that are based on Joseph Schumpeter's viewpoints, theories, and studies on the growth of innovation and entrepreneurship. Using a survey design and Yemane sample size determination formulas, a valid sample of 257 participants had an impact on the study. Intriguingly, the matrix analysis showed a substantial correlation between innovativeness and creative thinking, as well as a negative correlation between entrepreneurship development and technological advancement. This suggests that the development of technology supports creativity and innovation. Its direct impact on the growth of entrepreneurship, however, was not considered to be very important. To discover the truth, this can be looked into further. As a result, this is a plea for policy framework and support for all nations aspiring to economic achievement, not just Nigeria.

2.3 Muslim and hijab fashion worldwide

The word "hijab" is literally means to veil, to cover or to screen ([1], p. 476). According to Ismail Raji' al-Faruqi, Hijab means the covering of a Muslim woman or the Islamic style of dressing for women ([33], p. 28). The English word "scarf" and the Arabic term "hijab" have been used interchangeably, but the hijab has cultural connotations in many Muslim countries. Mohamood [34] illustrates that traditional and contemporary Islamic scholars and sociologists consider the veil as an assurance for the integrity of women and a protection from being used as a sexual object by predators. The term "hijab" in this study refers to the headscarf or veil that covers the hair of Muslim women when they are dressed appropriately and modestly. Muslim women show their identity and spiritual faith by donning a headscarf as a cultural and religious ritual [35]. Thus, there are inner and spiritual meanings for wearing the veil in the physical realm. It only portrays physically a lesson that must be understood spiritually [35].

In recent years, the hijab has emerged as a sign of Islamic consciousness, as more women are considering wearing the hijab as a symbolic of their conviction and dedication to be part of an Islamic Resurgence [36–50] especially in countries where wearing it is not compulsory or in non-Muslim countries [51] ([1], p. 477).

It is now possible to redefine the Muslim sign as a result of the hijab's growing popularity. The popularity of "Muslim women who dress stylish and/or make fashionable garments, while orienting towards what is being proposed by their religion in terms of attire" (hijabists or hijabistas), also known as "Muslim women who wear the hijab," has contributed to the emergence of the hijab fashion ([52], p. 161).

As a result of the growth of Muslim fashion media, bloggers, designers, and merchants, Muslim women today have more options for Muslim clothes, designs, and styles than ever before [53]. Siti Hasnah and Harmimi [1] summarised Wilson's [54] conclusion that the hijab fashion trend went as far as offering loose clothing by bridging communities locally through producing hijab fashion that stick with Muslim cultural terms and extending these offerings globally to eagerly awaiting Muslims consumers around the world. The current hijab fashion trends make an effort to update a modest look while preserving the fundamentals of traditional and cultural beliefs. This fashion produces a new style and trend for Muslim women who are fusing fashion with their Islamic faith, giving off a revitalising contemporary feeling of elegance and vigour.

As a transformational socialisation tool for Malay-Muslims in Malaysia's heterogeneous society, Nurzihan Hassim [55] explored the hijab, or the act of veiling. Using images of veiled Muslim women who are primarily Malay, the hijab was presented in the media as a progressive life style. Aside from Islamic upheavals and the globalisation of veiling in Muslim-majority nations, local socio-political changes have commodified the hijab as a Malay sovereignty-produced item rather than a representation of religious responsibility. Malay-Muslim women are more likely to be exposed to media messages that challenge traditional Malay-Muslim stereotypes as a result of the fusion of old and new media channels. The study investigates the local hijab phenomenon by analysing media coverage of it and its impact on viewers' attitudes and behaviours. Findings show that accessibility and diversity of content resonate with acceptance of this dynamic cultural identity.

Siti Hasnah and Harmimi [1] explore the factors influence the fashion consciousness in hijab fashion consumption among Muslims. The predictors of hijab fashion consciousness and consumption has been examined. Muslim women in developing countries, in particular the Muslim countries such as Malaysia whereby the number of educated, employed women has increased, and the country has changed and progressed from a traditional to a modern lifestyle. They have changed into contemporary, affluent Muslim women who adhere to the Shariah's dress code by dressing modestly and donning the headscarf. Because of this, the hijab fashion among "Muslim women who wear fashionable attire with matching fashionable headscarves" known as "hijabistas" is thriving.

Utilising the convenience sample method, data were collected by distributing questionnaires to Muslim women who attended the 2014 Kuala Lumpur International Hijab Fashion Fair. SmartPLS was utilised to analyse a total of 345 final usable data. The results show that fashion consciousness is positively influenced by dressing style, fashion motivation, fashion uniqueness, and fashion knowledge sources, which in turn adversely impact hijab fashion consumption which in line with the Shariah-compliant dress code [1]. Even though, Siti Hasmah dan Harmimi ([1], p. 490) have revealed their important outcome based on the conceptual framework, but the study should not ignore the important construct for Muslim products consumption, that is religiosity or religious values in the instrument of hijab consumption. This has been realised by both of them in their limitation and suggestion for future research statement: "In terms of construct, future studies must focus on other constructs, such as religious values or religiosity, in the Hijabista market segment".

Due to pandemic restrictions, foot traffic at retail establishments has decreased, which has negatively impacted modest fashion as well as the rest of the fashion business ([56], p. 5). Sales have been saved, nonetheless, thanks to the shift towards multichannel marketing and e-Commerce. The modest luxury fashion company Elle B Zhou used a 3D virtual representation of a Saudi model for its introduction, and the Turkish modest fashion powerhouse *Modanisa* introduced a new network in Malaysia with an online fashion show. Regularly through partnerships, mainstream clothing companies have continued to introduce their own lines of modest clothing. Muslim fashion spending expanded by 5.7% in 2021, from USD$279 billion to USD$295 billion, and is projected to expand by 6.0% in 2022 to USD$313 billion and USD$375 billion in 2025, for a 6.1% for a four-year Compound Annual Growth Rate (CAGR) [56].

2.4 Shariah compliance in the Muslim dress

According to Ismail Raji' al-Faruqi, ([33], pp. 36–37), al-Shari'ah means the collection name for all the laws of Islam, including Islam's whole religious and liturgical, ethical and jurisprudential systems. Shariah is an Arabic word which means the path that should be followed by Muslims [44]. Moreover, Shariah also means the totality of Allah's commands which regulate life for every Muslim in all aspects of life [57].

Syariah compliance is a term used to describe the practise of adhering to the laws obtained from the Qur'an, the Prophet Muhammad's Sunnah, agreement, analogy, and other acceptable and approved sources derived from these beginnings. These sources are used by the legal expert to determine which laws were most likely intended by the lawgiver [58]. In regard to the general guidelines in Islamic fashion for Muslim women, Hanisa et al. [59] describes that there are about 500 verses from the Quran and more in the Hadith regarding women's dress that should be followed by the Muslim women. Allah SWT mentioned in one of these verses as below.

> *"O Prophet, enjoin your wives and your daughters and the believing women, to draw a part of their outer coverings around them.[110] It is likelier that they will be recognised and not molested.[111] Allah is Most Forgiving, Most Merciful."*
>
> *(Al-Quran, Chapter Al-Ahzaab, 33: verse 59).*

In general, the regulation written in the Quran and the Hadith, are meant to protect women from harm [59]. Some of the major guidelines can be summarised in **Table 1**.

Islam has never prescribed a particular design style for Muslims to obey, despite the rules stated in the holy book of the Quran and illustrated with numerous examples in the Hadith. Despite being very generic, the command contains detailed instructions for Muslims to abide with. Therefore, Muslim women have the option to interpret the guidelines in light of their culture, environment, or activity, as long as it adheres to the instructions provided [59].

Both the Qur'an and the Sunnah place a strong emphasis on maintaining modesty when it comes to attire and style. Many popular aspects of Muslim dress, such as wearing longer-length clothing with more body coverage, opaque fabrics, and looser fits, reflect this ([56], p. 11). These still stand as some of the key characteristics that define what constitutes a garment as "modest" among Muslims around the world, despite the varied definitions of "modesty." Muslim nations are placing an increased emphasis on employing halal products and sustainable and ethical manufacturing processes [56].

Entrepreneurship within the framework of the Shari'ah compliance presents the impact that the Islamic religion has in the areas of realm of operation, financing, the forms of business, innovation, consumerism, women entrepreneurship, and social interest [60]. In the case of Muslim hijab fashion industry, the element of creativity and innovation in each piece of art and creative design must be complied with the Shariah principles.

Fundamental guidelines	Rationale of the principles
Garment must cover the whole body from unpermitted man (*non-mahram*)*	In general, every guideline instructs Muslim women to avoid drawing attention to themselves by exposing any part of their bodies. Although not all guys have a bad desire for women, the advice is meant to prevent it.
Dress must not be transparent or semi transparent	
Dress must not be tight fitting that will resembles the body shape	
Not to use strong perfume until it will draw attention of others	
Not to show off their jewellery or even make indication of it (like sound from a bell) as it may led to larceny.	
Not to dress like non-believer (by not covering their aurat4. They must not display act to attract men.	A Muslim woman must differentiate themselves from the non-believer by dressing according to the guidelines; but still not losing their grace as woman.
Dress should not resemble as men	
Garment must not to be a form of showing honour/ wealth	Men and women should be humble by not showing off their material comfort or being arrogant.

*Source: Hanisa et al. ([59], p. 455).*Non-Mahrams or Ghayr Mahrams are any males that a woman is allowed to marry (such as a cousin or any random Muslim male), as well as any males that are currently prohibited from marriage but may one day be allowed to wed if certain conditions alter. To put it another way, a Muslim woman who is previously married is not allowed to marry another Muslim man while she is still married. She may, however, marry another Muslim man after divorcing her current spouse and completing the "Iddah" (waiting time following divorce). Source: http://www.islamicinformation.net/2008/07/mahram-in-islam-explained.ht ml (29.10.13).*

Table 1.
Fundamental guidelines in Islamic fashion for Muslim women.

2.5 The concept of the middle class

Being affluent, educated and agent of social change, the middle class gain prominence by their income level, and become a pressure group in struggling to the issues pertaining social change and socio-economic development [61], ([62], p. 131), [63]. This study uses a combination of occupational categories and household income to define the middle class based on the most recent measurement of social class status from a Malaysian perspective. According to the Tenth Malaysian Plan 2011–2015 (2010), the middle 40% of Malaysians' household income in 2009 ranged between RM 2300 to RM 5599, while the top 20% of Malaysians' household income was greater than or equal to RM 5600.

Wan Nor Azriyati et al. [64], categorise the lower-income category includes individuals who earn less than RM1000, the lower-middle income category includes those who earn between RM1,501 and RM2,000, the middle-middle category includes those who earn between RM2,001 and RM4,000, the upper-middle income category includes those who earn between RM4,001 and RM6,000, and the high income group includes [64]. The household income range outlined in the Government Transformation Plan 2011–2015 document and Wan Nor Azriyati et al. [64], as well as the occupational model of Abdul Rahman Embong [65], are all modified as a guide for this study [61]. Moreover, Malaysia Standard Classification of Occupations 2008 [66] has classified 10 classification of major occupations as follow (**Table 2**).

From these occupational model and household income range, the middle class can be divided into two groups: those who earn between RM 2301 and RM 5599, and those who earn between RM 5600 and RM 10,000 or more and are considered upper-middle class. According to the aforementioned classification, professionals, technical, administrative, management, clerical, and service sector individuals who earn a minimum monthly household income of RM 2300 to RM 10,000 or more are considered middle class [61]. In the context of the current study, the Malay women entrepreneurs have been categorised with these range of occupation and household income.

Group	Occupations
1	Managers
2	Professionals
3	Technicians And Associate Professionals
4	Clerical Support Workers
5	Service And Sales Workers
6	Skilled Agricultural, Forestry, Livestock And Fishery Workers
7	Craft And Related Trades Workers
8	Plant And Machine Operators And Assemblers
9	Elementary Occupations
0	Armed Forces

Source: Malaysia Standard Classification of Occupations 2008, 2019. Available from: https://www.jobsmalaysia.gov.my/emasco/pengenalan-emasco#home.

Table 2.
Major occupations in Malaysia standard classification of occupations 2008 [66].

3. Methodology

This study employs a qualitative approach by using interview techniques or oral history through purposive sampling. The sample was selected based on the criterion sampling. A total of ten (10) informants who reside and were involved in the study. Data were analysed descriptively using content and thematic analysis. In addition, observation methods are also utilised to strengthen data collection of the study.

3.1 Research design

This study is primarily conducted based on exploratory qualitative methodology for this multidisciplinary of sociology and religious study. Documentation of experiences and knowledge through the semistructured in-depth interviews with 10 informants of the Malay middle income class group of women entrepreneurs was conducted using interview or oral history technique. The informants were chosen from a group individuals who are pioneers, who started their active involvement in Islamic activities in the 1970s in Selangor. Being pioneers in the Muslimah fashion industry, they share common thoughts and experiences with regard to the resurgence of Muslim and hijab fashion industry in Malaysia.

Interview sessions were conducted with selected representatives of the Malay women entrepreneurs in the group of the upper-middle and middle-middle income classes to obtain their insights and opinions on creativity and innovations and its relations to Shariah compliance. A semi-structured interviewing protocol was created to investigate the traits, behaviours and views of the Malay women entrepreneurs in Selangor, Malaysia. Instead of relying on the earlier literature, the semi-structured interview process encouraged the informants to emphasise the topics they thought were most crucial. According to Adams [67], semi-structured interviews are ideal for a variety of useful activities, especially when several of the open-ended questions call for follow-up questions. Adams [67], then proposes to consider using SSIs in the following circumstances in particular:

- If you want to learn about each person's autonomous opinions in a group and you need to ask probing, open-ended inquiries.

- If you want to use probing, open-ended questions to elicit information on subjects that focus group participants might not be willing to discuss openly.

- If you want to carry out a formative programme evaluation and want private interviews with important programme administrators, employees, and front-line service providers.

- If you are looking at unexplored region with unknown but potentially important issues, and your interviewees require the most freedom to identify and investigate useful leads.

From the transcripts of the interviews, the researcher extracted themes or dimensions. Descriptive and content analysis were used to examine the developed themes and dimensions. Thematic analysis (TA) which is proposed by Braun and Clarke [68], which later revised to their contemporary approach, known as Reflective Thematic Analysis (RTA) [69–73] was used to guide this study.

3.2 The selection of informants

Sample in this qualitative research was selected purposefully guided by a set of criteria. This research employed purposeful sampling for the reason that it would be able to understand the phenomenon of Muslim fashion industry and how it could influence on the Malay women entrepreneurs in Selangor to participate in this industry. The strategy for purposeful selection of information-rich cases is based on a specific type of purposeful sampling, that is criterion sampling proposed by Patton [74]; [61, 74]. According to Patton [74], the purpose of criterion sampling is to pick all cases that meet some criteria. Selecting examples with lots of information allows for deeper analysis, which is the rationale and power of deliberate sampling. The term "purposeful sampling" refers to selecting examples with the intention of learning as much as possible about topics that are essential to the investigation's goal. Instead of producing empirical generalisations, studying "information-rich cases" or situations which produces insights and thorough knowledge ([74], p. 169). In the context of the current study, the samples chosen fit the set of criteria.

Three criteria were used to justify the sampling strategy used in this study. First, a sample was picked using a categorical selection method, with Malay women entrepreneurs from the middle class as the representative group. Second, Selangor is home to many successful Malay women business owners. Third, the Malay women business owners chosen for the study are those who worked in Selangor's Islamic and Muslimah fashion industries.

In order to calculate the sample size, the principle of data saturation was applied, and new participants were added until further data stopped to provide fundamentally novel and unique insights. According to Mwita [75], when a researcher realises that they have gathered all the necessary data and that there is no longer any fresh, pertinent information or data that can be obtained from the study's respondents or subjects, this is referred to as saturation [75, 76]. Saturation typically means that a researcher needs to stop gathering new data for a certain topic. Data saturation has been widely employed in social science research and has emerged as one of the key components of the qualitative approach, which has its origins in the Glaser and Strauss grounded theory (1967) ([75], p. 414). Therefore, in the context of the current study, at a sample size of 10, data saturation occurred.

3.3 Data analysis

Meaning data was collected through the interviews. The average duration of each interview session was two and a half hours, and with the informants' consent, the interview sessions were audio-taped in their entirety and transcribed verbatim resulting in 345 pages of data. The transcription of the interviews was then converted into a Microsoft Word document.

The transcripts were analysed by categorising the text according to a six-phase process proposed by Braun and Clarke [69], which are described in **Table 3**. Thematic Analysis (TA) is utilised to analyse and combine massive amounts of data from naturalistic settings into insightful descriptions [68]. TA provides a technique for finding patterns or themes of meaning in a dataset and for interpreting and understanding their significance [69].

Thematic Analysis (TA) principles were manually applied to the data in order to find patterns of meaning that underlay the transcripts. Before categories were created and given names, this included a very iterative process including comparative analysis of patterns that emerged from the data.

Phase	Examples of procedure for each step
1. Familiarising oneself with the data	Transcribing data; reading and re-reading; noting down initial codes
2. Generating initial codes	Coding interesting features of the data in a systematic fashion across the data-set, collating data relevant to each code
3. Searching for the themes	Collating codes into potential themes, gathering all data relevant to each potential theme
4. Involved reviewing the themes	Checking if the themes work in relation to the coded extracts and the entire data-set: generate a thematic 'map'
5. Defining and naming themes	On-going analysis to refine the specifics of each theme; generation of clear names for each theme
6. Producing the report	Final opportunity for analysis selecting appropriate extracts; discussion of the analysis; relate back to research question or literature: produce report

Source: Szedlak et al. [77].

Table 3.
Six-phase thematic analysis procedures by Braun and Clake [68].

4. Findings and analysis

This study involves 10 Malaysian Malay women entrepreneurs as the informants. A brief background of the informants is as follows:

As stated in **Table 4**, that the Malay women entrepreneurs from the group of upper-middle and middle-middle income class informants who were involved in the Muslim fashion industry are the adults between the ages of 20 to 45 years old. This means that the informants' year of birth is between 1973 and 1998. The informants also lived in main urban cities of Selangor, namely Shah Alam, Bangi, Klang, Kajang, Selayang and Puchong, Petaling Jaya.

The Malay women middle income class entrepreneurs were chosen from diverse background of education. Albeit having considerably good academic qualifications in different disciplines, they have shown relatively similar convictions to religious commitment in implementing Shariah principles in their dress and fashion designs. In fact, their educational background does not limit their quest to exhibit significant manifestations of their faith in thoughts and practices as well as demonstrate their religious responsibility through the Shariah-compliant products. Most of the informants demonstrated significant adherence to the religion of Islam and thus resulting in high commitment to religious obligation in which their prepare more choices in the Islamic dressing design and fashion for Muslims either in the national and global market.

The following discussion is divided into several sub-themes, as to highlight the elements of creativity, innovation and the Shariah compliance in the Muslim fashion industry and how Islam has impacted the lives of the Malaysian Malay women entrepreneurs in their profession as key players of Islamic fashion industry in Malaysia.

4.1 The factors that led to the involvement of women entrepreneurs in the Muslim fashion industry

Various factors have been contributed to the involvement of the Malay Muslim women entrepreneurs in the urban state of Selangor in the Muslim and hijab fashion

Identity	Position in the Muslim fashion company	Level of education	Age	Place of living
Informant 1 – I1	Manager	Degree	41	Shah Alam
Informant 2 – I2	Supervisor	Diploma	28	Kajang
Informant 3 – I3	Founder	Degree	30	Shah Alam
Informant 4 – I4	Founder	Degree	27	Shah Alam
Informant 5 – I5	Co-founder	Diploma	20	Puchong, Petaling Jaya
Informant 6 – I6	Founder and Manager	Diploma	22	Selayang
Informant 7 – I7	Co-founder	UPSR	30	Klang
Informant 8 – I8	Supervisor	SPM	45	Shah Alam
Informant 9 – I9	Founder	Degree	29	Bangi
Informant 10 – I10	Founder	Degree	29	Bangi

Table 4.
The background of the Malay middle income class informants.

industry. Based on the transcriptions of the interview sessions with the informants, five thematic factors have been grounded and emerged. The factors are religious responsibility for the sake of the Ummah; self interest in entrepreneurship, family support, positive attitude towards success and improving soft skills for survival. The emerging factors could be described as the followings:

4.1.1 Religious responsibility for the sake of the Ummah

The conviction towards religious responsibility has been the driving force of their career undertakings. Some of Malay women entrepreneurs put preference on their convictions before their career. This has been due to many reasons, such as the serve Muslim with *wudu'*-friendly and maternity designs which has led to religious commitments. Such religious awareness is intrinsic in nature.

> *"The products supplied and produced must be coincided with the Islamic law. For me, the Muslimah fashion products must comply with the Shariah compliance itself in order to avoid defamation from God" (I1).*
>
> *"As a Muslimah fashion designer and entrepreneur, the element of creativity and innovation should be harmonized with the Shariah-compliant.... We try to convince our customers that we will always produce the best for the adherents of the Muslimah fashion who always put their confident on our collections" (I2).*
>
> *"I produce the designs that the Arabic-style robe (jubah) that are wudhu'-friendly and I do not use materials that could expose and fit tightly to human body, like lycra fabric. And other products that I have produced are for example, the "A-cut Muslimah t-shirt". This design would be loose and comfortable for big size individuals" (I6).*
>
> *"Actually, at the beginning, I am really not interested to start a business, but the world has changed. I am now happy and very grateful to thank Allah SWT [syukur], and Alhamdulillah, because now, I really love, admire and happy and even deeply interest in my own business that promotes the Muslim fashion products to serve for the sake of Ummah. This is my jihad and da'wah in economics" (I7).*

> *"Our designs are produced on the costumer design basis. They may suggest their own design exclusively and we offer an advice to the designs that still preserve the Shariah principles, especially covering the aurah of the customer" (I9).*
>
> *"As an entrepreneur, this is my responsibility to offer a collection of garments that certified with the Shariah-compliant verification as well as stylish recent fashion. I think that the basic aspect of the Shariah such as covering aurah must be considered in advance" (I9).*
>
> *"Any creativity effort in designing a product need to put a priority on the aspect of aurah which should be covered properly. For that reason, I choose fabrics that do not cling to the customers' body as well as offers the best design for the customers" (I10).*

4.1.2 Self interest in entrepreneurship

Entrepreneurial interest or inclination is described as motivating factor that affects the individual to pursue in an entrepreneurial venture [78]. According to Bardai [79], the absence of deep interest for running a business will lead to the failure in entrepreneurship. This is proven by the description given by Arifin and Sabaruddin [80], who emphasised that a high interest can encourage and inspire someone to be persistent when dealing with risk.

Moreover, positive interest in a particular field of endeavour is the key to one's success. Individual interest enable someone strives towards something desirable. Otherwise, lack of interest would influence the determination of a person and thus would degrade the performance of its achievements in a particular field. The results showed that all informants have deep interest in entrepreneurship and business.

> *"Most people who start a business in order to seek the money. Money is a leading motivating factor to start a business, but I did a business due to my personal interest. Self-interest has motivated me to continue studying in this field to ensure that my business could be succeeded and lead to the progress in the coming years" (I6).*
>
> *"Actually, at the beginning, I am really not interested to start a business, but the world has changed. I am now happy and very grateful to thank Allah SWT [syukur], and Alhamdulillah, because now, I really love, admire and happy and even deeply interest in my own business that promotes the Muslim fashion products to serve for the sake of Ummah. This is my jihad and da'wah in economics" (I7).*
>
> *"I have intensely interested in business since primary school to find my own pocket money for the reason that my family who lived in a simple life request e to work hard for survival" (I3).*
>
> *"I really have a deep interest in the Muslim cloth design, so I do not really look for the profit solely, because I am more concern about customer satisfaction and acceptance" (I4).*
>
> *"A deep interest and dedication in the world of business had been contributed to my serious involvement in this arena despite many challenges that I faced" (I5).*

4.1.3 Family support

Some informants describe that family support and encouragement help them to develop their interest in business. Family is not just a focal factor of their success in business, but also become a backbone that always back them up when they face tribulations and problems in business.

> *"My husband is a very supportive mate of my life. He continues to support me at home with my children. We do many activities together, playing and studying with children. My husband helped me to create an online business and manage to run online marketing to increase sales and purchases" (I1).*
>
> *"My father and my sister also had involved in business. I have experienced and started to study business since my childhood because I always follow my father in managing our family business" (I2).*
>
> *"I manage to begin my business since I was studying at the university. At that time, I did not have enough capital and investment to start the business. Fortunately, I have my family that had provided me motivational and financial supports to start my new business and starting from that I managed to involve in entrepreneurship of the Muslim fashion" (I4).*

The informants' convictions about the role of the family which contribute to their involvement in business have been explained by many researchers. Parents have crucial and significant roles to generate and inculcate interest in their children based on their free wills and choices. The previous studies show that parental influence on learning helps to encourage and inculcate good values and even this interest in individuals did not exist at the first place [81]. Moreover, Desforges and Abouchaar [82], also suggests that positive attitudes of parents and their encouragement would help to increase interest in children and improve their achievement, particularly in academic and school performance.

4.1.4 Positive attitude towards success

Most of the informants also show their positive attitude and enthusiasm towards success in business as well as throughout their life at the first place. Such positive attitudes for example, responsibility, positive thinking, put an effort to help the family in term of finance, improve social mobility and standard of living, working independently without hoping for others' assistance, being self-confident and self-reliant. These were shown through some of their confessions as follows:

> *"As the eldest in the family, I feel responsible to help my family as well as this could increase our social mobility and standards of living in term of economic. For example, we need to have a high income and to have a big car and big house" (I3).*
>
> *"I prefer to carry out my works without depending on other assistance. This is because I feel free to make any decision by myself" (I4).*
>
> *"Not everyone likes to be instructed by someone else and that is why I decided to stop working in the previous company. I'm a person that do not like to work under an authoritarian leadership and I want to do work that I really love. In other words, I want a freedom to do things that I like and do not involve other people who could give the instruction on me" (I5).*

Attitude is an abstract value that shape in the mind of the individual. It can create the perception in the individual [83]. To develop the interest of the individual in entrepreneurship and business, Mohammed Idris [84] advocates that a person's beliefs must build his or confident that he or she could succeed in this field. This optimistic attitude towards success is a very significant step that determine the future prospect of someone's life.

In addition, Mohd Osman [83] had listed some attitudes that are needed to be a successful entrepreneur. These are self-reliance, perseverance, creativity, innovation, receiving the challenges, hope and ambition, to see and grasp the opportunities, self-confident in individual's ability, energetic, enthusiasm, hardworking and good effort.

Even in the case of disabilities students, the parents and school should educate their students to build their positive attitude. For example, U.S. Department of Education [85] had addressed that the attitude and self-advocacy skills of students with disabilities may be two of the most important factors in determining their success or failure in postsecondary education. To ensure that students with disabilities possess the desired levels of self-advocacy to succeed in postsecondary education, high school educators may want to encourage the students to: First, understand their disabilities. Second, accepting responsibility for success, Third, take an appropriate preparatory curriculum. Fourth, acquire computer skills. Fifth, consider supplemental postsecondary education preparatory programs. Sixth, research postsecondary education programs. Seventh, get involved on campus.

4.1.5 Improving soft skills for survival

In addition to the previous factors, the informants' involvement in entrepreneurship has been triggered by their effort to improve the soft skills for survival. The followings are some of the informants' reasons for conducting business in the Muslim fashion industry.

> *"I have sewing skills and I realize that these skills can help me to earn more money and gain more profits. At first, I just sewed clothes as free time activities. Furthermore, I began to receive customers' orders for tailoring from nearest neighbors and friends. From there, I started to seriously involve in business, and I had decided to lunch a Muslim dressing and gauze boutique" (I2).*

> *"At the beginning, I manage a business through social media, such as blog. From there, I managed to develop social relationship with other people who eventually become customers, agents and suppliers via online business without physical face to face communication" (I4).*

> *"I am a friendly and verbose person. I like to ask other opinions, so that I can practice effective communication when dealing with the customers, the competitors and suppliers. Such skills are significant in business for the purpose of improving the quality of the products, as well as to gain better profits for both parties, the owner and the customers" (I5).*

The ability to use the internet and social media is the pre-requisite soft skills to market goods or product worldwide. Local entrepreneurs also need to think and act creatively in generating and stimulating unique ideas for their product. This includes to demonstrate the physical and non-physical products which could be the assets for business [86]. For example, the physical product can be viewed from the production, packaging, and branding, whereas the non-physical aspects can be viewed as managing a network of social communication with customers, family, followers and society in general. Thus, someone who has knowledge, the ability and expert in designing a business via social media communication has advantage to market their products very fast through this medium.

The findings of the previous studies also demonstrated the similar results that network relationships are important to entrepreneurs as they provide access to resources that are an important component in the entrepreneurial process [87]. In addition, seizing information is a very important skill for entrepreneurs in generating a market, making decisions, or solving problems [88]. The benefits of networking will engage these women entrepreneurs in entrepreneurial activities such as scanning, for, recognising and exploiting business opportunities [89]. This networking as the acquired soft skills could motivate Malay Muslim women entrepreneurs to involve in the Muslim fashion industry worldwide.

4.2 The importance of creativity and innovation in the production of the Muslimah fashion among the Malay Muslim women entrepreneurs of the middle class in Selangor

Most of the informants viewed that creativity and innovation are significant aspects in the production of any design of the Muslim fashion. These two elements are interconnected and crucial to attract customers and publicise the products either at the national and global market.

> *"Creativity and innovation in fashion are the factors that could attract customers to buy our products. If there are no such both creativity and innovation, the collection is not in line with the trendy fashion that being implemented by other competitors. Both are important in Muslim fashion due to customers' demand, and they always request the different and unique designs for daily and occasion dress" (I1).*
>
> *"The creativity and innovation are interrelated elements. Creativity can be described as a simple thing that attract others. Creativity also should be applied by fashion designers or people who involve in the Muslim fashion industry. They always have ideas in producing a product. So, from such creation and unique ideas it could improve the innovation of the product to be accepted in the market. The element of creativity and innovation are important in the creation of dressing and hijab designs" (I2).*
>
> *"Both creativity and innovation are the most important elements in the production of fashion design. The creativity in the collection could help to increase selling and retail in which these are very much dependent on the design that has been produced" (I3).*
>
> *"The creativity is an important part of fashion. Innovation is a process of implementing creativity in the different way in line with the latest design and trends. With the creative design, it could attract customers to buy the product" (I4).*
>
> *"Creativity is a talent that has been given by Allah SWT. It can also be learnt. I have no basic knowledge about fashion and have never studied in the university or any other learning institution. But I have learnt Muslim fashion from personal experiences. This element is very important in the Muslim collection, and this could be the reason why we promoted our products" (I5).*

4.3 The implementation of the Shariah compliance in designing creative and innovation Muslimah fashion design

Almost all informants understand their obligation to apply the Shariah-compliant aspects in designing any creative and innovative design for Muslim customers. This indicates that they realise on their religious responsibility to provide and promote a proper Shariah-compliant design for the Muslim fashion such as in dress, garments and head scarf. There are significant relationship between creativity, innovation and the Shariah-compliance in designing the Muslim fashion that can be proved through the informants' convictions.

> *"Shariah compliance is an Islamic principle of law. The Islamic law is primarily based on The Quran and as-Sunnah. The products supplied and produced must be coincided with the Islamic law. For me, the Muslimah fashion products must comply with the Shariah compliance itself in order to avoid defamation from God" (I1).*

> *"As a Muslimah fashion designer and entrepreneur, the element of creativity and innovation should be harmonized with the Shariah-compliant. These are imperative pre-requisite to our products in order to avoid any perception that the product is contradicted to the Islamic principles and is an obsoleted design. We try to convince our customers that we will always produce the best for the adherents of the Muslimah fashion who always put their confident on our collections" (I2).*

> *"Cover the aurah is the most important thing of Shariah compliance in Muslimah fashion design, which is not too excessive to the principle of Shariah" (I3).*

> *"Shariah compliance in Muslimah fashion is about covering the aurah that includes the private parts of women's body, lowering the hijab or head scarf until it cover the chest. Shariah compliance is a vital aspect in the Muslimah fashion" (I4).*

> *"The Shariah-compliant designs were characterized by covering the aurah of man and woman, the cloth must be loosed and not very tight or body-hugging until it would reveal someone's skin, and the dress is suitable to be wore by all people" (I5).*

> *"These three elements [creativity, innovation and Shariah Compliance should always be related one another. For example, how the producing ideas can be stimulated? Does it is aligned with what Islam teach? This means that when we produce a product, the material and design must be in line with the Islamic teaching" (I6).*

> *"Using the creative thinking, it can inculcate and instill the Islamic values in Muslim fashion" (I7).*

> *"From my understanding, they [creativity, innovation] are very closely interrelated to each other. Actually, most of our collection are abide by the Islamic principles. The products can be wore and suited to all level of society, elegant to the current trends and style with the sophisticated multi-cultural Islamic creative designs" (I8).*

> *"Our designs are produced on the costumer design basis. They may suggest their own design exclusively and we offer an advice to the designs that still preserve the Shariah principles, especially covering the aurah of the customer" (I9).*

> *"We combined the element of creativity, innovation and Shariah compliant in our collections and designs. For instance, we managed to use the up-to-date design, suggesting the new trend and at the same time we also emphasize the elements that are complied to the Shariah" (I10).*

5. Discussion

This study analyses the element of creativity, innovation and Shariah compliance in Muslimah fashion among Malay women entrepreneurs in Selangor, an urban state of Malaysia. The findings were categorised by the three themes that emerged during the interviews. Each of the findings justified the contributing factors to the Malay entrepreneurs involvement in the Muslim fashion; the importance of creativity and innovation in the Muslim fashion; and the implementation of the Shariah compliance in the Muslim fashion among the Malay entrepreneurs in Selangor.

The findings indicate that five thematic emerging factors have contributed to the participation of the Malay women entrepreneurs in the Muslim fashion industry in Selangor. The study shows that the majority of the Malay entrepreneurs venture into entrepreneurship due to first, religious responsibility for the sake of the Ummah; second, self interest in entrepreneurship; third, family support; fourth, positive attitude towards success; fifth, improving soft skills for survival. These findings are in accordance to the previous studies that revealed on the various factors that lead to the involvement of women entrepreneurs in business [90–92]. The factors are for example, interest, ambition to become an entrepreneur, seeks for job satisfaction and desire to create a working environment suite to their personality. The finding of the study is also consistent with the previous studies that reveal that the combination of entrepreneurial factors has a significant relationship with the success of an entrepreneur. Skill, knowledge, and adaptability in business are the main reasons for women to emerge into business ventures [93].

The findings also indicated that the role of family in particular, the parents are important social segment to inculcate good values and interest in children. The parents are responsible to provide not just the basic physical needs of their children, such as clothing, food and shelter, but most importantly, they need to give more attention to spiritual and religious needs in which these are the crucial facets in the individual development. Many studies have also proved that parental involvement and engagement in the early childhood, education and in making the decision, gave the positive impacts to their children' success in life that shapes their children future prospect [94–101].

According to the report conducted by the Global Entrepreneurship Monitor (GEM) 2015, the involvement of women in entrepreneurship area are not only contribute for the development of country, but also reduce the level of poverty all over the world [102]. In fact, the involvement of women in entrepreneurship is very important in order to mobilise the economic model whereby it creates more opportunities which is more competitive and profitable [103]. In the Malaysian context, women entrepreneurs enter the business world due family commitment (for rural women & single mothers), personal achievement, independency, and autonomy (for corporate women) which are parallel to the pull theory [89].

All informants of the Malaysian Malay women entrepreneurs understand the concept of creativity, innovation and perceive the Shariah compliance in their designs and collections. They considered and took into account the Shariah compliant as the priority in their designs which signify and reflect their religious obligation and commitment. The conviction to Islam has been the driving force of their career undertakings. Some of Malay women entrepreneurs put preference on their convictions before their profession in business.

Today, creativity and innovation appear more significant than ever before, with these being seen as critical success factors for organisations and companies [104]. For every modern business to succeed in a world marked by competition, technological change, and recurrent crises of post COVID-19 outbreak, innovation has become essential [6]. The understanding of attitudes towards creativity and innovation and promoting of creative thinking within the organisation are pre-requisites to facilitate creativity in all employees [105, 106]. To increase the value of entrepreneurship, for example, creativity and innovation have become essential. In order to improve

DOI: http://dx.doi.org/10.5772/intechopen.109655

current business practices, creativity is a great asset. Making a procedure that can turn an idea into a more saleable product or service is referred to innovation [14]. Therefore, the need and requirement for creativity and innovation are a key approach in entrepreneurship.

The Muslim fashion industry is rapidly growing worldwide. The sales volumes are raising, and the Muslim fashion companies have been significantly expanded their business empire to the Southeast East Asia, the United States, the Middle East and North Africa regions. Today, Muslim consumers spend a lot of money on dressing as a symbol of faith. The State of the Global Islamic Economy Report 2015/16 conveyed that in comparison to the current combined textile markets of the UK (USD$107 billion), Germany (USD$99 billion), and India (USD$96 billion), Muslim consumers are forecasted to spend USD$230 billion on clothes, with a projected increase to USD$327 billion by 2019 [107].

Muslim fashion consumers' and designers' decisions about clothing or other fashion products are influenced by religion [108] and culture [109–111]. While young Muslim women in Malaysia are currently expressing interest in Western design trends, they must also uphold Muslim principles about suitable wear. Therefore, they require reliable resources for fashion tips that reduce the social danger of consuming something illegal or for guidance on this new trend among Muslim consumers.

Based on the findings revealed by Aliman et al. [108], despite the hijab fashion culture being quite widespread in multicultural countries like Malaysia, the majority of Malay Muslim women who wear them do not consider them to be fashionable. They view wearing the hijab as a religious obligation rather than a fashion choice. In order to further strengthen Muslim religious identity and cultural diversity, the hijab is thus used as a powerful tool.

6. Conclusion

The Malay urban women entrepreneurs in the study show their dedicated religious commitment through Islamic branding of Muslim fashion. As described by Nurzihan ([112]: 428) that the role of Malay-Muslim women in creating a renewed perception upon the hijab in the Malaysian public, pursuing a more liberated, Islamic identity whilst offering a renewed social construction of the Malay society. This renewed social construction could be viewed from the glance of media through the evolution of mass communication such as in print, broadcast and social networking channels.

Acknowledgements

The authors would like to thank The Ministry of Higher Education, Universiti Sains Islam Malaysia for the financial support provided for this research.

Author details

Mashitah Sulaiman[1*] and Mohd Nazir Ahmad[2]

1 Center for Core Studies and Faculty of Leadership and Management, Universiti Sains Islam Malaysia, Nilai, Negeri Sembilan, Malaysia

2 Faculty of Information Management, MARA University of Technology, Shah Alam, Selangor, Malaysia

*Address all correspondence to: mashitah@usim.edu.my

References

[1] Siti Hasnah H, Harmimi H. Factors influencing fashion consciousness in hijab fashion consumption among hijabistas. Journal of Islamic Marketing. 2016;7(4):476-494. DOI: 10.1108/JIMA-10-2014-0064

[2] Farhad D. Hijab Fashion Industries 'Unveiled'. n.d. DOI: 10.1108/JIMA-10-2014-0064. Available from: https://www.ia-forum.org/Files/JISVAJ.pdf. https://www.ia-forum.org

[3] Brown J. How Muslim Women's Fashion is Shaping Conversation around Creativity and Power. 3 Jan 2019

[4] Juliana NO, Hui HJ, Clement M, Solomon EN, Elvis OK. The impact of creativity and innovation on entrepreneurship development: Evidence from Nigeria. Open Journal of Business and Management. 2021;**9**:1743-1770. DOI: 10.4236/ojbm.2021.94095

[5] Gholami K, Karimi A. The role of creativity, innovation and entrepreneurship in the organization. Bulletin of Environment, Pharmacology and Life Sciences. 2014;**3**(Spl issue II):73-77. https://bepls.com/vol3_spl_II/12.pdf

[6] Adam NA, Alarifi G. Innovation practices for survival of small and medium enterprises (SMEs) in the COVID-19 times: The role of external support. Journal of Innovation and Entrepreneurship. 2021;**10**(15):1-22. DOI: 10.1186/s13731-021-00156-6

[7] Runco MA, Jaeger GJ. The standard definition of creativity. Creativity Research Journal. 2012;**24**(1):92-96. DOI: 10.1080/10400419.2012.650092

[8] Gerlovina Z. Eureka! Unraveling The Mystery Behind Creativity. B.A. Thesis. Columbia University; 2011. Available from: https://academiccommons.columbia.edu/doi/10.7916/D8ZP4D3F

[9] George JM, Zhou J. Dual tuning in a supportive context: Joint contributions of positive mood, negative mood, and supervisory behaviors to employee creativity. Academy of Management Journal. 2007;**50**:605-622. DOI: 10.5465/AMJ.2007.25525934

[10] To ML, Ashkanasy NM, Rowe PA. Dual-Moods and Creativity in the Organisation: A Bidirectional Mood Regulation Perspective. 2008. Available from: https://pdfs.semanticscholar.org/5a6a/6ab8c2af75582d50c11104609df4f406a4ac.pdf

[11] Safra JE, Yeshua I. The New Encyclopaedia Britannica. Vol. 3. London: Encyclopaedia Britiannica Inc; 2002

[12] Nakano TC, Wechsler SM. Creativity and innovation: Skills for the 21st century. Estudos de Psicologia (Campinas). 2018;**35**(3):237-246. DOI: 10.1590/1982-02752018000300002

[13] O'Hara M. Rising to the occasion: New persons for new times. Estudos de Psicologia (Campinas). 2017;**34**(4):454-466. DOI: 10.1590/1982027 52017000400002

[14] Muhamad Izzuwan Shah Bin Mazla, Mohd Khata Bin Jabor, Kashif Tufail, Amir Faisal Noor Yakim, Hanim Zainal. The roles of creativity and innovation in entrepreneurship. Advances in Social Science, Education and Humanities Research. 2020;**470**:213-217

[15] Goh BK. Creativity, Innovation and Entrepreneurs. Strategy and

Innovation Management in Tourism and Hospitality Industry. n.d. Available from: http://www.trm.chu.edu.tw/ezfiles/74/1074/img/868/LEC5-INNOVATIONANDENTREPRENEURS.pdf

[16] Ceserani J, Greatwood P. Innovation & Creativity: Getting Ideas—Developing Solutions—Winning Commitment. London: Kogan Puge Limited; 1996

[17] Porter M. What is strategy? Harvard Business Review. 1996;**74**(6):61-78

[18] Tornatzky LG, Fleischer M, Chakrabarti AK. Processes of Technological Innovation. Lexington Books. 1990

[19] Alsuwaidi AKMS, Alshami SA, Suriati Akmal. The impact of entrepreneurship towards innovation In airport industry: The double mediation framework of strategic alignment and learning orientation. Academy of Strategic Management Journal. 2021;**20**(2):1-19

[20] Kim DY, Kumar V, Kumar U. Relationship between quality management practices and innovation. Journal of Operations Management. 2012;**304**:295-315

[21] Carayannis E, Grigoroudis E. 'Linking innovation, productivity, and competitiveness: Implications for policy and practice'. Journal of Technology Transfer. 2015;**39**:199-218

[22] Fillis I, Rentschler R. The role of creativity in entrepreneurship. Journal of Enterprising Culture. 2010;**18**(01):49-81

[23] Amabile T. A model of creativity and innovation in organisations. In: Staw BM, Cummings LL, editors. Research in Organisational Behaviour. USA, Greenwich, CT: JAI Press; 1988. pp. 123-167

[24] Ortiz-Villajos JM. Patents, what for? The case of Crossley brothers and the introduction of the gas engine into Spain, c. 1870-1914. Business History. 2014;**56**(4):650-676. DOI: 10.1080/00076791.2013.837890

[25] Gaynor G. Innovation by design: What it takes to keep your company on the cutting edge. Amacom. 2002

[26] Bilton C. Management and Creativity: From Creative Industries to Creative Management. Oxford: Blackwell Publishing; 2007

[27] Petrowski MJ. Creativity research: Implications for teaching, learning and thinking. Reference Services Review. 2000;**28**(4):304-312

[28] Oke A, Munshi N, Walumbwa FO. The influence of leadership on innovation processes and activities, Organizational Dynamics. 2009;**38**(1):64-72

[29] Van Den Broeck H, Cools E, Maenhout T. A case study of arteconomy: Building bridges between art and enterprise: Belgian businesses stimulate creativity and innovation through art, Journal of Management and Organization. 2008;**14**:573-587

[30] McMullan JS, Shepherd DA. Entrepreneurial action and the role of uncertainty in the theory of the entrepreneur. Academy of Management Review. 2006;**31**(1):132-152

[31] Ballor JJ, Claar VV. Creativity, innovation, and the historicity of entrepreneurship. Journal of Entrepreneurship and Public Policy. 2019;**8**:513-522. DOI: 10.1108/JEPP-03-2019-0016

[32] Adzmi MH, Ishak Z, Ladin CA. Creativity, design and the approaches to technology. International Journal of

Advanced Research in Technology and Innovation. 2022;**4**(1):9-17. Available from: http://myjms.mohe.gov.my/index.php/ijarti

[33] Ismail Raji' al-Faruqi. Toward Islamic English. Virginia, USA: The International Institute of Islamic Thought. 1986

[34] Mahmood S. Politics of Piety: The Islamic Revival and the Feminist Subject. Princeton: Princeton University Press. 2005

[35] Galadari A. Behind the veil: Inner meanings of women's Islamic dress code. International Journal of Interdisciplinary Social Sciences. 2012;**6**(11):115-125

[36] Mohamad AB. Without Locke or Lenin: The Impact of Resurgent Islam on International Politics. Kuala Lumpur: Bahagian Komunikasi Korporat, Universiti Malaya; 2005

[37] Huntington SP. The Clash of Civilizations and the Remaking of World Order. New York: Simon & Schuster; 1996

[38] Mutalib H. Islam and Ethnicity in Malay Politics. Oxford: Oxford University Press; 1990

[39] Mutalib H. Islamic revivalism in ASEAN states: Political implications. Asian Survey. 1990;**30**(9):877891

[40] Mutalib H. Islam in Malaysia: From Revivalism to Islamic State? Singapore: Singapore University Press; 1993

[41] Mohamad AB. Islamic revivalism and the political process in Malaysia. Asian Survey. 1981;**21**(10):1040-1059

[42] Othman MR. Merentas Sempadan yang Melewati Zaman: Timur Tengah, Islam dan Masyarakat Melayu. Syarahan Perdana. Kuala Lumpur: Dewan Kuliah A, Fakulti Sastera dan Sains Sosial, Universiti Malaya; 2015

[43] Nagata J. Religious ideology and social change: The Islamic revival in Malaysia. Pacific Affairs. 1980;**53**(3):405-439

[44] Nagata J. The Reflowering of Malaysian Islam: Modern Religious Radicals and their Roots. Vancouver: The University of British Columbia Press; 1984

[45] Nagata J. How to be Islamic without being an Islamic state. In: Ahmed AS, Hastings D, editors. Islam, Globalization and Postmodernity. London: Routledge; 1994

[46] Muzaffar C. Islamic Resurgence in Malaysia. Petaling Jaya: Fajar Bakti; 1987

[47] Dessouki AEH. Islamic Resurgence in the Arab World. New York: Praeger; 1982

[48] Esposito JL. (ed.) Voices of Resurgent Islam. Oxford: Oxford University Press. 1983

[49] Hamayotsu K. Islam and nation building in Southeast Asia: Malaysia and Indonesia in comparative perspective. Pacific Affairs. 2002;**75**(3):353-375

[50] Ahmad K. The nature of islamic resurgence. In: John L. Esposito (ed.). Voices of Resurgent Islam. Oxford: Oxford University Press. 1983

[51] Stacey A. Why Muslim Women Wear the Veil. 2009. Available from: www.islamreligion.com/articles/2770/why-muslim-women-wear-veil/ [Accessed: July 1, 2015]

[52] Blommaert J, Varis P. Culture as accent: The cultural logic of Hijabistas. Semiotica. 2015;**203**:153-177

[53] Ghani AA. Asian Muslim Women's Fashion History. Aquila; 2011

[54] Wilson JA. Ramadan reflections. The Marketeers. 2015:115-118

[55] Nurzihan H. Glocalizing the hijab: A Malaysian perspective. SHS Web of Conferences 33, International Conference on Communication and Media: An International Communication Association Regional Conference (i-COME'16), Department of Media Studies, Faculty of Arts and Social Sciences, Universiti Malaya, Malaysia. 2017

[56] The 2022 State of the Global Islamic Economy Report. Dubai: DinarStandard Growth Strategy Research and Advisory. 2022

[57] Hayes DK, Ninemeir JD. Hotel Operation Management. 2nd ed. New Jersey: Pearson Prentice Hall; 2007

[58] Oz E, Ali MM, Zahid I, Khokher Z, et al. Shari'ah Non-Compliance Risk in the Banking Sector: Impact on Capital Adequacy Framework of Islamic Banks. 2016. Available from: https://www.ifsb.org/docs/2016-03-30SNCRPaper(WP-05)(Final).pdf

[59] Hanisa H, Zaman BA, Santosa I. Tolerance of Islam: A study on fashion among modern and professional Malay women in Malaysia. International Journal of Social Science and Humanity. 2015;**5**(5):454-460

[60] Ghoul WA. Entrepreneurship within the framework of Shari'ah. International Journal of Business and Globalisation. 2015;**15**(3). DOI: 10.1504/IJBG.2015.071904

[61] Sulaiman M. Islamic Resurgence and Religiosity among the Malay Middle and Low Income Classes in Selangor and Kuala Lumpur, 1977-2014. PhD. Thesis. Faculty of Arts and Social Sciences, University of Malaya; 2016

[62] Ahmad MN, Sulaiman M, Zakaria S, Mutalib MMA, Samsudin AZH. Documenting knowledge through Oral history: Islamic awareness among the Malaysian Malays. In: BE-ci 2016 : 3rd International Conference on Business and Economics. Malaysia: UiTM, Shah Alam Selangor; 2016. pp. 131-146

[63] Coleman R, Rainwater L. Social Standing in America: New Dimensions of Class. New York: Basic Books, Inc. 1978

[64] Wan Nor Azriyati WAA, Hanif NR, Singaravello K. Affordable housing within the middle-income households in Malaysia: Challenge to enter homeownership. Australian Journal of Basic and Applied Sciences. 2011;**5**(8):263-264

[65] Embong AR. State-Led Modernization and the New Middle Class in Malaysia. New York: Palgrave; 2002

[66] MASCO. Ministry of Human Resources Malaysia (2010). Malaysia Standard Classification of Occupations 2008 (MASCO 2008). 3rd edition. Putrajaya: Ministry of Human Resources Malaysia. 2008. pp. 1-286

[67] Adams WC. Conducting semi-structured interviews. In Newcorner, Kathryn E, Hatry HP, Wholey JS. Handbook of Practical Program Evaluation. Jossey Y-Bass. 2022

[68] Braun V, Clarke V. Using thematic analysis in psychology. Qualitative Research in Psychology. 2006;**3**(2):77-101. DOI: 10.1191/1478088706qp063oa

[69] Braun V, Clarke V, Weate P. Using thematic analysis in sport and exercise research. In: Smith B, Sparkes AC, editors. Routledge Handbook of Qualitative Research in Sport and Exercise. London: Routledge; 2016. pp. 191-205

[70] Braun V, Clarke V. Successful Qualitative Research: A Practical Guide

for Beginners. Sage Publications, Thousand Oaks. 2013

[71] Braun V, Clarke V. Reflecting on reflexive thematic analysis. Qualitative Research Sport Exercise Health. 2019;**11**(4):589-597. DOI: 10.1080/2159676X.2019.1628806

[72] Braun V, Clarke V. One size fits all? What counts as quality practice in (reflexive) thematic analysis? Qualitative Research Psychology. 2020. DOI: 10.1080/14780887.2020.1769238

[73] Byrne D. A worked example of Braun and Clarke's approach to reflexive thematic analysis. Qualitative & Quantitative. 2022;**56**:1391-1412. DOI: 10.1007/s11135-021-01182-y

[74] Patton MQ. Qualitative Evaluation and Research Methods. 2nd ed. Newbury Park, CA & London: Sage; 1990. Available from: https://legacy.oise.utoronto.ca/research/field-centres/ross/ctl1014/Patton1990.pdf

[75] Mwita KM. Factors influencing data saturation in qualitative studies. International Journal of Research in Business & Social Science. 2022;**11**(4):414-420

[76] Fusch P, Fusch G, Ness L. Denzin's paradigm shift: Revisiting triangulation in qualitative research. Journal of Social Change. 2018;**10**(1):19-32

[77] Szedlak, Chistoph, Smith, Matthew J., Day, Melissa C., and Greenlees, Iain A. (2015). Effective Behaviours of strength and conditioning coaches as perceived by athletes. International Journal of Sports Science & Coaching. 10(5), pp. 966-984.

[78] Hisrich RD, Peter MP, Shephard DA. Entrepreneurship (International Edition). Singapore: The McGraw-Hill Education Asia; 2008

[79] Bardai B. Keusahawanan dan Perniagaan. Kuala Lumpur: Dewan Bahasa dan Pustaka; 2000

[80] Arifin S, Sabaruddin AS. Keusahawanan: Rahsia ke Puncak Kejayaan. Kuala Lumpur: Prentice Hall and Pearson Malaysia Sd. Bhd; 2002

[81] Durisic M, Bunijevac M. Parental Involvement as a Important Factor for Successful Education. Center for Educational Policy Studies Journal. 2017;7(3):137-153. Available from: https://files.eric.ed.gov/fulltext/EJ1156936.pdf

[82] Desforges C, Abouchaar A. The Impact of Parental Involvement, Parental Support and Family Education on Pupil Achievement and Adjustment: A Literature Review. Research Report No 433. 2003:1-108. Available from: https://www.nationalnumeracy.org.uk/sites/default/files/documents/impact_of_parental_involvement/the_impact_of_parental_involvement.pdf

[83] Mohd Osman MH. Faktor-faktor yang telah Mendorong Graduan Alumni UTM Menceburi Bidang Keusahawanan. B.A. Thesis. Universiti Teknologi Malaysia; 2007

[84] Mohamed Idris SH. Kecenderungan Keusahawanan di Kalangan Pelajar Bidang Kejuruteraan di Institut Pengajian Tinggi Awam di Kawasan Utara Semenanjung Malaysia. M.A. Thesis. Universiti Utara Malaysia; 2009

[85] U.S. Department of Education, Office for Civil Rights. Keys to Success: Attitude, Self-Advocacy and Preparation. In: Transition of Students with Disabilities to Postsecondary Education: A Guide for High School Educators. Washington, D.C.: U.S. Department of Education, Office for Civil Rights; 2007

[86] Nurdin G, Chan GK, Selvadurai S, Ishak S. Hubungan sosial dan p e r n i a

g a a n kecil-kecilan di Malaysia-Tingkah laku inovatif usahawan wanita atas talian. Geografia: Malaysian Journal of Society and Space. 2014;**10**(6):206-218

[87] Rajennd A, Muniady L, Al Mamun A, Permarupan PY, Zainol NRB, Mohamad MR. Social capital: Cross-industry and cross-state comparison among Malaysian women Micro-entrepreneurs. Mediterranean Journal of Social Sciences. 2016;**7**(1):350-360

[88] Makhbul ZM, Hasun FM. Entrepreneurial success: An exploratory study among entrepreneurs. International Journal of Business and Management. 2011;**6**(1):116

[89] Xavier SR, Ahmad SZ, Nor LM, Yusof M. Women entrepreneurs: Making a change from employment to small and medium business ownership. Procedia Economics and Finance. 2012;**4**:321-334

[90] Alam S, Jani M, Omar N. An empirical study of success factors of women entrepreneurs in southern region in Malaysia. International Journal of Economics and Finance. 2011;**3**(2):166-175

[91] Brush CG. Research on women business owners: Past trends, a new perspective and future directions. Entrepreneurship Theory and Practice. 1992;**16**(4):5-30

[92] Moore DP, Buttner EH. Women Entrepreneurs: Moving beyond the Glass Ceiling. Thousand Oaks: Sage Publications; 1997

[93] Alagappar PN, Vaithilingam K, Hussain FC, Tien WYM, Xavier JA. A research overview of women Entrepreneur's in Malaysia: How far have they come? Journal of Holistic Student Development. 2017:1-10

[94] McWayne C, Melzi G. Family engagement in children's preschool experiences among low-income Latino caregivers: The validation of a culture-contextualized measure. Journal of Family Psychology. 2014;**28**:260-266. DOI: 10.1037/a0036167

[95] Melendez MC, Melendez NB. The influence of parental attachment on the college adjustment of white, black, and Latina/Hispanic women: A cross-cultural study. Journal of College Student Development. 2010;**51**(4):419-435

[96] Moon UJ, Hofferth SL. Parental involvement, child effort, and the development of immigrant boys' and girls' reading and mathematics skills: A latent difference score growth model. Learning and Individual Differences. 2016;**47**:136-144

[97] Moore EK. Increasing Parental Involvement as a Means of Improving Our Nation's Schools. Washington, D.C.: National Black Child Development Institute; 1990. Available from: http://files.eric.ed.gov/fulltext/ED325232.pdf

[98] Newchurch A. The impact of parental involvement on student success: School and family partnership from the perspective of parents and teachers. In: Doctor of Education in Teacher Leadership Dissertations. The Bagwell College of Education, Kennesaw State University; 2017

[99] Piper T. Using school website to support parent engagement. Leadership. 2012;**42**(2):36-38

[100] Rafiq HM, Fatima T, Sohail MM, Saleem M, Khan MA. Parental involvement and academic achievement: A study on secondary school students of Lahore, Pakistan. International Journal of Humanities and Social Science. 2013;**3**(8):209-223

[101] Redding S, Langdon J, Meyer J, Sheley P. The Effects of Comprehensive

Parent Engagement on Student Learning Outcomes. San Diego, California: Presented at the American Educational Research Association Convention; 2004

[102] Rani SHA, Hashim N. Factors that influence women entrepreneurial success in Malaysia: A conceptual framework. International Journal of Research in Business Studies and Management. 2017;**4**(1):16-23

[103] Malaysia. Jabatan Perdana Menteri. Unit Perancangan Ekonomi. The Tenth Malaysia Plan 2011-2015. 2010. Putrajaya: Unit Perancangan Ekonomi, JPM

[104] Basadur M, Hausdorf PA. Measuring divergent thinking attitudes related to creative problem solving and innovation management. Creativity Research Journal. 1996;**9**(1):21-32

[105] Agor WH. How intuition can be used to enhance creativity in organizations. Journal of Creative Behaviour. 1991;**25**(1):11-19

[106] Mason RB. The external environment's effect on management and strategy: A complexity theory approach. Management Decision. 2017;**45**(1):10-28

[107] Mucci A. Inside the Booming Muslim Fashion Industry: Muslim Fashion Designers Reflect on a Rapidly Growing Industry. 2016. Available from: https://www.aljazeera.com/indepth/features/2016/01/booming-muslim-fashion-industry-160124132747636.html

[108] Nurzihan H. Hijab and the Malay-Muslim woman in media. The international conference on communication and media 2014 (i-COME'14), 18-20 October 2014, Langkawi, Malaysia. Procedia—Social and Behavioral Sciences. 2015;**155**(2014):428-433

[109] Almila AM. Fashion, anti-fashion, non-fashion and symbolic capital: The uses of dress among muslim minorities in Finland'. Fashion Theory. 2016;**20**(1):81-102

[110] El-Bassiouny Noha. The Hijabi self: Authenticity and transformation in the Hijab fashion phenomenon. Journal of Islamic Marketing; Bingley. 2018;**9**(2):296-304. DOI: 10.1108/JIMA-12-2016-0102

[111] Zain Mohd Zaimmudin Mohd. The Influence of Muslim Fashion Bloggers on Young Muslim Females' Fashion and Life in Malaysia. Ph.D Thesis. The Faculty of Science and Engineering, The University of Manchester. 2020

[112] Nurzihan Hassim. Hijab and the Malay-Muslim woman in media. The international conference on communication and media 2014 (i-COME'14), 18-20 October 2014, Langkawi, Malaysia. Procedia—Social and Behavioral Sciences. 2015. 155 (2014). pp. 428-433

Chapter 7

A Worldwide Regression Analysis of Relations between Creativity, Innovation, and Quality Management Culture, under a Psychological Approach

Enriko Ceko

Abstract

This chapter aims to demonstrate the significant connections between innovation and creativity, innovation and the culture of quality management, and creativity and the culture of quality management globally as a response to the Crisis and Post-Crisis Period. Since issues related to innovation, creativity, quality, and quality management have recently attracted more attention than other topics globally, considering psychological factors was of great importance, too. The methodology of the research involved collecting data and information about innovation, creativity, and ISO standards application globally, in higher-income, medium-income developing, and low-income developing countries. The main finding of this study is that there are relationships between innovation and creativity, between innovation and ISO standards application, and between quality management and creativity. The main conclusion of the study is that quality management, which reflects and refers to the notion of exceeding customer expectations by enhancing the quality of goods and services, is correlated with culture, innovation, and creativity as a response to the Crisis and Post-Crisis Period.

Keywords: creativity, innovation, quality management, quality culture, factors of production

1. Introduction

In the neoclassical theory, land, labor, and capital were considered, but capital is divided into fixed capital, labor capital, and financial capital. Regarding the factors of production in the classical theory (Adam Smith, etc.) of economic studies, the factors of production were considered land, labor, and capital. In the Marxist theory (Karl Marx, Frederik Engels, etc.), it was considered labor, the subjects of work, and the means of work. In environmental economics, production factors were known beginning in the latter half of the twentieth century, in addition to matter, energy, and design intelligence, the entrepreneurial spirit started to be discussed as a factor in production (entrepreneurship).

While this chapter emphasizes that innovation is a quality shared by employees as well as entrepreneurs, some academics regard innovation as a component of entrepreneurship. Accordingly, for innovation to occur, subjects or companies must establish the framework in which employees can carry it out.

Creativity is a prerequisite for innovation and cannot happen without it. When there are enough and the right conditions, an individual can move from creativity to innovation. This person is distinguished by their creative abilities. If the invention is sound, the market will experience a competitive and comparative economic advantage, which is further aided by ongoing quality improvement and the adherence to and use of quality management systems and standards.

During the pandemic period, when the traditional view of factors of production where labor, land, and capital were treated as the most important factors is almost finished, as well as discussing only entrepreneurship, since entrepreneurship is the most recent factor of production, the combination of early factors of production (land, labor, and capital) with modern factors of production (entrepreneurship and technological innovation) made it possible for many entities operating in different sectors of the global economy to succeed [1].

For this reason, the components of products that were also taken into account in this study include labor, land, capital, entrepreneurship, and innovation.

The past economic crises, particularly the one from 2007 to 2013, natural disasters with a focus, pandemics, particularly the COVID-19 virus, have all increased the value of a culture of quality management, innovative activities, and creative ideas, which, when combined, have positive economic effects.

Inventions, as well as new ideas and concepts, were already associated with creativity. Currently, creativity receives more attention across several fields. This attention was mostly focused on the social sciences, humanities, economics, entrepreneurship, health, agro-processing, education, engineering, philosophy, and so on, and is primarily focused on ideas produced by the actions of both individuals and groups of individuals [2].

Typically, innovation is defined as the combination of new features that leads to meliorated and/or newly discovered goods, innovative processes, procedures, regulations, record keeping, manipulations, manufacturing, improved penetration in existing markets, introduction to new markets or clients, actualization of new models of resource usage, meliorated business models, and so on.

In most cases, this is closely related to how creativity and innovation are supported, the effectiveness of entrepreneurial activities (processes, procedures, regulations, orders, products, services, methodologies and methods, tools, technologies, etc.), the human resources involved in these creative and innovative processes, as well as the processes of quality and its management. The resulting goods and services are marketed to clients as standalone goods and services or in combinations, primarily through B2B, B2C, and B2G. These products are one-of-a-kind shoes of the highest caliber at reasonable pricing.

This enables these products to enter the market more quickly and with lower marketing expenses. Innovation does not need to result in an invention. It is sufficient to have a product or service with a more practical application approach. Making decisions to address issues related to enhancing the quality of life is today the primary objective, and this unquestionably involves innovation. Therefore, new, novelty, and innovation all refer to the creativity that is shown in both individual and group activities.

It makes sense that a business would need to make quality and cultural investments to attain, maintain, and manage creative and innovative operations.

The values of the previous work's accomplishments and successes, as well as its ongoing improvement, comprise the quality culture. It is precisely this approach to culture in general and to the culture of quality in particular that makes it possible, through interventions that correct and prevent low quality, to make it possible to offer quality goods on the market at the most suitable prices for economic agents. It is precisely this culture that makes it possible to improve the quality of the products and services provided by various entities.

Successful businesses around the world appear to have had and now have a focus on "doing the job well the first time." This is demonstrated by (1) the development of human resources both individually and within work groups, (2) the establishment of tolerance and respect as the cornerstones of organizational communication activities, (3) managerial responsibilities and procedures, and (4) entrepreneurial activity, as fundamental, main values that unquestionably have a positive impact on the economy.

They are precisely ISO standardized, which technically embodies the quality culture. Both the public and private sectors have shown an increasing interest in ISO standards in recent years, particularly during the pandemic. This was because these standards were primarily concerned with the improvement of processes, procedures, regulations, and data retention, which enables the further improvement of product and service quality to meet consumer expectations.

The incorporation of managerial functions, such as planning, budgeting, leadership, motivation, and control, into an organization's day-to-day operations enables it to further define and improve its quality policies, objectives, and responsibilities. These managerial functions can be strategic, operational, or applied, and they enable an organization to put these policies, objectives, and responsibilities into practice through ongoing improvements in the caliber of its products and services. This occurs every day. This is precisely what ISO standardization and quality management system are for.

2. Literature review

The reason why the literature on quality and its management, creativity and creative abilities, innovation and novelty, and so on, continues to advance, expand with new knowledge, and develop more and more on a scientific basis is that these concepts when applied with care and intelligence, enable all types of public and private entities to achieve, maintain, and manage economic success. This phenomenon of the global economy existed for at least the last 20 years of the twentieth century and the twenty first century in markets where supply is much more than demand.

2.1 Main 2020–2030 megatrends and how the world has responded to the crisis and the crisis period and later too

The following megatrends will dominate the next 10 years:

- Economic growth associated with the shift of economic power.
- Significant changes in business practices and investment strategies are anticipated in the economy of developing nations.
- The yield of agricultural and livestock goods will be impacted by global warming, which will result in changes in food product pricing. Communities lacking enough financial resources will be impacted by this.
- Artificial intelligence, machine learning, and digital (r)evolution will be buzzwords of the day.
- Compared to the US, Asia has a larger population over 65.
- As this segment of the population continues to grow, there will be more persons over 65 in Asia by 2042 than in the Eurozone and North America put together.
- Social changes, which present both challenges and opportunities for governments and human society, will come before changes in the world's demographics.
- It is these tendencies specifically that have brought and will continue to bring significant, widespread, and quick changes and advancements in technology, structure, and superstructures, bringing undeniable influences and effects on international markets and societies that are increasingly global.
- At now, metropolitan regions are home to more than 50% of the world's population. There will probably be 5 billion of them by the year 2050, according to estimates. The African continent, Asia, and so on, will see a significant portion of this urbanization process, which will result in significant political, economic, social, cultural, legal, environmental, and other transformations in various regions of the world.

As a result, humans are heading in the following directions in the face of these quantitative and qualitative changes, which are the megatrends that will influence every part of humanity's lives for at least the next 50 years:

- Information and technology have undergone a revolution.
- Increasing and broadening of the flexibility of both public and private organizations that are looking to adopt innovative technologies and processes.
- The development and growth of knowledge, competencies, and learning gained from market competition.
- Increasing worldwide offer and competitiveness.
- The use of a variety of tactics and settings by commercial entities.
- Pay attention to the knowledge economy.

- Systematic upgrades to manufacturing, distribution, and redistribution.
- A greater requirement for the integration of policies.
- An increase in global output and investment.
- Shifts in the economy's structure.
- Investment in and development of competencies and knowledge.
- The application and growth of systems for creativity, innovation, and quality management.

2.2 Factors of production

The inputs utilized in the production process for financial gain are included or described as factors of production. As previously noted, the factors of production in this essay are as follows: (1) land, (2) labor, (3) capital, (4) innovation, and (5) entrepreneurship. (Although some authors include entrepreneurship innovation, innovation is not a sole attribute of entrepreneurs, but of employees and other individuals who are not involved in business activities as well).

It is clear, the cyclical nature of the economy and the cyclical use of production factors have been and continue to be closely related to world history regarding factors of production since preparing the paper carefully followed the developments of the global economy for at least the last 40 years and read the literature on the history of the development of the global economy for at least the last 600 years.

The use of physical factors—land, labor, and capital—that have always been regarded as fixed factors of production was 1 cycle that is currently understood. The other cycle is the use and management of creativity, innovation, and entrepreneurial skills, knowledge, and competencies—that are regarded as non-fixed factors of production. It has been the cycle of use, exploitation, and management of land, labor, and capital that has predominated in some historical and economic development periods, while it has been the cycle of use and management of skills, knowledge, and competencies related to creativity, innovation, and entrepreneurship in other historical and economic development periods. The stages of applying these components, which are the stages of entry, growth, maturity, and decline, are what lead to all of this. One cycle enters the entry phase if it is in the decline phase, and the other cycle enters the endpoint phase if it is in the peak maturity phase.

Nowadays, there is a significant quantity of territory (land) in almost every nation that is not utilized, exploited, or managed. Unemployment is extremely high everywhere in the world and primarily affects young people. A lot of free money is also available around the globe, both in banks and in people's pockets, but it is not being invested.

Therefore, it is evident and simple to see that the cycle of use, exploitation, and management of land, labor, and capital is in decline or may have even reached the end of the decline phase. In contrast, the cycle of exploitation, use, and management of creativity, innovation, and entrepreneurial skills is rising or even at its peak. It is difficult to predict how long this phenomenon will last, during which the cycle of the use of fixed factors of production will be declining and the cycle of non-fixed factors of production will mature. However, one thing is certain: When 1 cycle reaches its growth peak, the other cycle reaches its endpoint, and vice versa.

One of the key components for increasing productivity and economic growth has been, is, and will continue to be the rapid technological advancement and the accumulation of skills, knowledge, and competencies; in this regard, systems and networks that effectively and efficiently disseminate information, skills, knowledge, and competencies are crucial. For this, policies about education, science, industries, and technologies as well as incentives for investments in research and development, ongoing training, support for systems of creativity, innovation, and quality management, as well as infrastructure for these purposes are necessary [3, 4].

2.3 Creativity

Years of quality management experience have shown that creative (1) processes, (2) procedures, (3) goods, (4) services, (5), and so on, are an important components of creativity or contain it [5–7]. This demonstrates, in essence, that creativity is what initiates first and establishes the conditions for inventive activities to occur, acting as a continuous mental and psychological process for problem identification and solution selection [8].

The process of creativity, which consists of several stages and is characterized by a thinking process where (1) fluid, (2) flexible, (3) original, and (4) elaborative ideas emerge in human brains, includes problem definition, data, figures, and information collection, selecting the best solution among several of them, and from this point creating improved or newly discovered products and services.

Organizations need management expertise and practices, motivation, knowledge, mental acuity, techniques, procedures, people with flexibility and imagination to approach issues, and internal access to the organization related to the elements external (e.g., threats to leave the job or income) and internal to achieve and maintain creativity (such as job satisfaction, employee development).

At this point, employee motivation practices include:

- Choosing the right challenge to present to individuals;
- Achieving objectives, creating conditions for people's autonomy to choose the best tools;
- Balancing human, financial, and material resources, as well as time;
- Creating an environment that supports working groups that support each other;
- Evaluation, supervision, and mercy;
- Support for organizations, sharing of information and knowledge, as well as collaboration [15].

A sign of entrepreneurship studies and the social and psychological skills, knowledge, and competencies of entrepreneurs is the ability of employees and businesspeople to use creativity to realize products and services with improved and new features. This is accomplished using processes of continuity, brainstorming, conceptions, support, and confidence [9].

According to some authors [10–12], creativity and knowledge creation are essential for successful organizations. These authors also emphasize how communication, infrastructure, technology, and training all play important roles in this process.

Parallel to this, the idea of the "creative class, an important driver of the economies of the modern era," in conjunction with the "3 T regions" (Technology development, talent empowerment, and tolerance for differences), leads to a high concentration of creativity among professionals who strive to have a higher position of economic development [13].

2.4 Innovation

According to an OECD report [14] and numerous authors [15–19], business is an innovative process that involves human, financial, and material resources as well as the capacity to produce, using skills, knowledge, and managerial and entrepreneurial competences that are developed during a long and important creative process, although at the beginning, the process can be when subjects submit their goods and services to the market, innovation appears to be a process of the practical implementation of ideas brought by the market, and when these subjects work diligently and consistently to enhance their goods and services. Innovation is defined as "an improvement or something new that presents and brings back value" by the ISO 56000:2020 Standard [20].

Improvement, invention, and dissemination—three components of innovation—are shared by all authors who have looked at various facets of this topic [21]. While innovation cannot be viewed as only an invention [22] or as if every innovation is an invention in and of itself [23], not all innovations necessitate dealing with the invention [24].

The creation, distribution, and application of skills, knowledge, and competencies within a system can be positively impacted by laws, procedures, and processes, but excessive bureaucracy and a plethora of restrictions are detrimental to innovation [25].

There are three stages to the innovation process: idea generation and idea empowerment, problem resolution and decision-making, and implementation [26].

The primary drivers of innovation are significant system failures, chance, and earnest efforts made by numerous economic agents. Along with this, there are the movements and demographic shifts, the shift in people's attitudes toward innovation and other topics, and the degree of knowledge, skills, and current scientific capabilities, which are important sources of innovation [27].

The tools and techniques used for problem-solving and decision-making, activities that include agents of many and different economics, and this constitutes a very important aspect regarding the outcomes of public and private entities, which are all included in this aspect and are therefore essential to the success, efficiency, and effectiveness of innovations.

The innovative aspects of products and services are brought to the attention of consumers by combining marketing with its components (advertising, publicity, public relations, and promotion), a process that involves communication, a process that includes the generation of ideas, the work of individual and group intellectuals, and so on, and mostly this mental and intellectual work, innovative technological changes rather than any activity that is necessarily and primarily related to the use of fixed factors of production, land, labor, and capital [28].

The connections between management, leadership, entrepreneurship, inspiration, and creativity have been studied by several authors [29–32].

According to certain writers, the Triple Helix model [33] serves as an infrastructure for innovation, which fosters interaction and the development of the knowledge economy [25, 34]. They contend that those engaged in science and academia are those who produce and disseminate knowledge. These are the facts that the market's economic agents gather and apply [33]. The company generates and offers funding and opportunities for the development of skills, knowledge, and competencies to support innovation, and the government creates and provides incentives to promote and foster innovation [25].

Every incident involving such things as pandemics and natural calamities fosters standard-setting, innovation, and creativity. The growing interest in innovation has been observed during the Covid-19 pandemic period in the areas and industries of public health, education, distant labor, electronic commerce, and so on [35].

2.5 Concepts of quality and quality culture

Quality, which is ultimately connected to consumers' expectations for the features of goods and services, demonstrates that it is also connected to customers' perceptions, which are conditioned and subjective. Since meeting customer demands at a particular level is essentially what customers desire, quality is defined by ISO standards as doing so.

It was Edward Deming who first suggested that cost-cutting must accompany productivity development to enhance quality. This approach to quality improvement calls for the use of administrative functions, proper processes, and procedures, as well as design and testing methodologies. Quality is what buyers search for in goods and services, according to Peter Drucker. Consumers are always willing to pay more for quality than for the money that producers wish to receive. This leads to the perspective that the essence of quality and quality culture is being able to satisfy consumer requests.

The relationship between quality, quality management, and quality culture and the organization's management, in general, has to do with the strategy of placing the customer at the heart of the business and attending to their demands—a long-term dedication to the client, collaboration, the kaizen approach (constant improvement), ongoing training for employees during their tenure with the organization, setting up the right conditions for the fulfillment of freedom of action through control, empowerment, and involvement of employees, and so on. This quality system encompasses factors such as employees, workplace health and safety, the manufacturing process, organizational finances, information creation and management, research and development, procurement practices [2, 36].

The relationship between management principles and ISO 9001 is depicted in the image below (**Figure 1**):

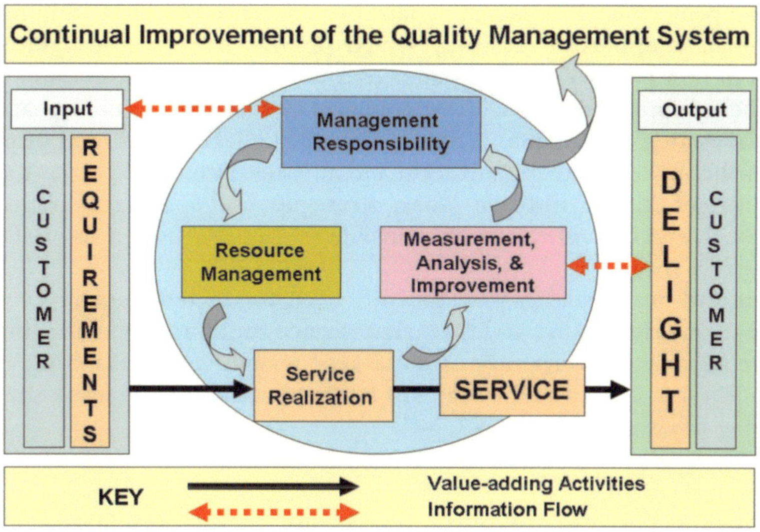

Figure 1.
The management principles and ISO 9001 relations [37].

The values necessary to realistically implement continuous improvement are related to quality. For this, several factors are required, through which it is possible to develop the organizational philosophy, the workgroup culture, the mobilization and motivation of people involved in creative and innovative projects, and so on. Many scholars in the field of quality management characterize this as a social conflict that keeps the organization's members united [38].

In the jargon of quality management, this is exactly what subject culture means; it is what ties individuals in an organization together and connects them to the values they support and develop, with how these people naturally use organizational structures, controls, and mechanisms, and so on [39]. This culture encourages you to strive for excellence and creates a setting where everyone in the company is committed to raising quality. Now, each employee sees himself as a customer, a supplier, and an employee. It's a great creative idea to strive to complete everything as well as you can right away.

First, from this perspective, it should be made clear that it is not claimed that since there is room for improvement, there is no reason to do everything well from the start, as this mentality increases costs for the organization. Instead, the approach aims to evaluate every step of the creation of products and services in every organization where a quality culture exists and is functional, as this prevents the passing of time [40].

Thus, the improvement of both individuals and organizations along the path of continuous realization of goods and services, the development of a climate of respect and tolerance among employees, and the development of entrepreneurial skills, knowledge, and competencies are all closely related to quality, its management, and quality culture (4). The capabilities of the organization and proof that these capabilities are regarded and used properly.

The domains that are known and valued by every individual who comprises the organization and who identifies themselves with these domains are what constitute the quality culture, which is manifested in the quality of products and services [41].

The main principles of "achieving goals through training" and "bringing benefits through value and opportunities" [42], which aim for excellence, recognition, and application of best practices and experiences to achieve and maintain standards, are precisely these approaches to quality and quality culture that set Europe apart. This is a managerial strategy that makes it possible for the beliefs, values, and principles of the people who make up the organization and all other interested parties to create and use the culture of this organization. Goals, strategies, and operational objectives are clearly defined, and individuals are responsible for predetermined and defined processes and procedures.

In this regard, many studies have recently been conducted on quality, quality management, quality culture, and their significance for attaining and maintaining competitive advantages, corporations' social responsibility, sustainability, and organizational continuity, as well as the ethics of business activity and raising standards of living, among other topics [2, 39, 43, 44].

2.6 Relationships between creativity, innovation, and quality management in psychological terms

Technology, business, social systems, economic development, and policy formulation are just a few of the contexts in which creativity, innovation, and quality management have been explored. As a result, there are many different ways to conceptualize creativity, innovation, and quality management, including the psychological approach.

The study of conscious and unconscious phenomena, such as feelings and thoughts, is included in psychology's scope as a discipline that examines the brain and human behavior. As a whole, psychology is a significant academic discipline with a broad scope that blurs the lines between the humanities and the natural sciences. While social scientists seek to comprehend the behavior and attitudes of individuals and their groups in specific situations, psychologists seek to comprehend the properties of the brain by relating the activity of the brain and its properties to neuroscience, in their daily lives and aspects of organizational life [45, 46].

In addition to health-related techniques, psychological knowledge is used to comprehend and resolve issues in a variety of human endeavors, including industrial and organizational settings [47].

I/O psychology is the study of, and application of, psychological theories and principles to the working environments of organizations [48]. Initially known as economic psychology or business psychology, industrial psychology, employment psychology, or psychotechnology [49], and emerging as I/O psychology between 1960 and 1970, industrialists applied the young field of I/O psychology to the study of scientific management techniques for enhancing workplace efficiency.

It became a part of Division 14 of the American Psychological Association's Society for Industrial and Organizational Psychology in 1973 [49]. The goal of this profession is to improve how people behave in work environments and procedures. Personnel psychology is one of I/O psychology's subfields. To recruit and assess employees, personnel psychologists specifically attempt to use psychological methodologies and concepts. Another branch of psychology is called organizational psychology, which looks at how the workplace and various management philosophies affect workers' productivity, job satisfaction, and motivation [50].

I/O psychology studies are currently concentrating on creativity, innovation, and quality management as well. The activities and studies related to this field (human engineering psychologists) are practically connected to the problem-solving and decision-making approach, such as how to organize work, choose workers, design machines that are easy for people to use, create and innovate new products/services or improve existing ones, and apply the updated standards, procedures, processes, and good practices aimed at competitive advantage achievement.

In fields as diverse as technology, health, and tourism, innovation is acknowledged as the cornerstone of development and wealth [51].

The discipline of psychology, which studies every aspect of human experience, has a favorable impact on creativity, innovation, and crisis management.

Information on the psychology of innovation discusses the traits of innovators and creative solutions to everyday issues that are grounded in psychological science. The promotion of innovative workplaces highlights the significance of setting expectations, defining metrics, and fostering a positive work environment for employees [52]. Other methods focus on how creativity and innovation interact within larger sociocultural contexts [53].

Some studies claim that to innovate, people must alter their behavior; therefore, in addition to qualities like creativity, it is also necessary to be able to encourage behavior change [54].

Understanding the distinction between original thought and actual innovation—thought that is helpful, implementable, and likely to have a significant impact on the field—is necessary if companies are to be vibrant, innovative, and productive [55].

The most intriguing definition of innovation is thinking differently [56]. People with creative skills engage in more diverse thinking, and they can produce thoughts and reactions that are different from those of ordinary people [57]. They think differently than other people do, and they approach problems differently as a result. They have the capacity for unusual thought and the ability to initiate dialog, promote debate, and present options that may not have previously been considered. Nowadays, according to many psychologists, an action must possess at least two qualities to be considered truly innovative.

It ought to be special and practical. An idea needs to be practical, solve issues, and benefit people in some way to be considered creative [58]. The application of creativity by the organization's staff is essential to developing an innovative culture. By using the adage "If you throw anything at a wall, you can better determine what goes on that wall," many ideas can be produced.

However, truly innovative solutions can bring a more accurate definition and redefinition of problems in the organization and/or open a completely new way of looking at its solution. Creativity can bring solutions to problems carried over and newly born in the organization by offering clear solutions and reconsidering the ways of doing business [59].

This means that the solutions that are ultimately found may not be what was initially regarded to be the best course of action. Innovative and creative ideas have the power to transform an organization and fundamentally alter the way that individuals and teams approach problems, look for answers, and reach choices. First, according to this perspective, effectiveness and efficiency are key components in the concept of innovation and the capacity for productivity.

It can inspire creativity in even the wise in a period of increased output. Simply challenging people to think creatively can boost their enthusiasm and encourage them to engage in innovative work. It has been observed that the existence of an innovative approach in an organization causes other individuals who have not previously engaged in innovation to rethink how they view issues and potential solutions, uncovering new approaches, resources, and possibilities. According to this perspective, in a workplace where innovation is present, employees contribute more ideas, many of which are unique and useful.

Divergent thinking can be learned through close contact with people who possess creativity and invention, according to some studies, which also suggests that creativity and innovation might be "contagious." [60].

In general, creativity is a part of innovation but cannot be confused with it because innovation has to do with creative ideas to bring certain and obvious improvements to particular products and services where innovation is necessary due to market need. Innovation typically starts with creative thought, and in this sense, creativity—whether it be individual or collective—brings innovation. Innovation is a prerequisite for creativity, but it is not a sufficient condition [61].

It takes more than just having an idea, coming up with one, or coming up with original ideas for innovation to occur. Knowledge is put to use to bring about changes through novel, distinctive methods, resulting in the introduction of new goods and services to the market for consumers. The fact that innovation, like many administrative and entrepreneurial responsibilities, necessitates particular methods, procedures, norms, record-keeping, and discipline is one of the most crucial facets of this topic [13].

With this strategy, the emphasis is now on the existence and ongoing development of processes, procedures, regulations, and record-keeping for the development and application of skills, knowledge, and competencies, which lead to significant positive organizational changes, to achieve, maintain, and manage improved products and services, and to improve in the near, medium, and distant future, as an added value of the organization's activity, which is essentially the organization's business.

Through these various perspectives, creativity is typically seen as the foundation for innovation, and innovation is the successful implementation of creative ideas within an organization [61]. However, to be competitive, organizations must use quality management principles and standardize their processes and procedures, and the goods and services they provide [37].

Researchers have now succeeded in defining several innovation types:

1. A shift in how the company is conducted in terms of creating and managing value.

2. The advancement of marketing techniques through novel approaches, such as product design and packaging.

3. Reorganization of organizational structures, functions, protocols, rules, and record-keeping practices.

4. Novel processes that use new or improved techniques to carry out manufacturing and service or deliver them to the end users.

5. Products that are largely or entirely innovative, including enhancements to technical and functional features, how goods and services are presented, and so on.

6. Cutting-edge services, which are comparable to product upgrades.

7. Innovative supply chains, where improvements are made to the quality, safety, and assurance of goods from the point of entry to the point of exit.

8. Financial innovation, which is demonstrated by the creation of new financial services and products.

The improvement of quality, the opening of new markets, the expansion of the variety of goods and services, the reduction of business costs, the improvement of processes, procedures, regulations, and record keeping, the reduction of the use of materials, energy, and water in technological processes, environmental protection, and compliance with laws and bylaws all seem to be related to investments in innovation today. All of these objectives are guided by the concept of continuous improvement, which is one of the core tenets of quality management.

Aspects of personality theory suggest that human personality is made up of five broad dimensions: openness, conscientiousness, extroversion, agreeableness, and neuroticism. Some personality traits are also linked to creativity, innovation, and quality management.

Each of these factors constitutes a segment where the person may receive a high, low, or indifferent rating for each component. Being open to new experiences is a crucial skill that is closely linked to flexibility, creativity, and quality management.

Having said that, those who exhibit this trait more strongly are more receptive to novel situations and concepts. However, other personality traits have a significant part in creativity, innovation, and business management. They tend to strive for innovation and find fulfillment in the fact that they try new things, meet new people, view problems from other angles than those they have previously encountered, and so on.

Intrinsic drive, curiosity, and perseverance, for instance, can all influence how much people are inclined to investigate unique concepts and look for inventive solutions while using established standards, procedures, and processes.

TQM may be explained by nine essential psychological processes, including identification, equity, equality, consensus, instrumentality, rationality, development, group dynamics, and internalization, according to many studies [62].

Total Quality Management is used by around 75% of Fortune 1000 organizations, according to research. The survey results indicate that 87% of respondents believe they have had a favorable experience with total quality management. In these situations, TQM is related to improving the financial condition, relationships with employees and among themselves, product and service quality, customer satisfaction levels, and so on. It was determined through research of 54 companies that use total quality management that their performance was better than that of groups of comparable organizations that do not use total quality management.

Companies that use these programs are more admired are seen as being more inventive, and have higher management quality ratings than other companies [63].

The qualities of innovation, creativity, and quality management are shown below (**Table 1**) along with their relationships.

Creativity, innovation, and quality management		
Techniques of creativity	Innovation	Quality management
1 Finding and respecting aim & approach;	1 Radical changes: "brings an innovative design that affects production and related components"	1 Total product/service features;
2 Forming abilities;		2 Quality management;
3 Supporting getting of new specific knowledge;		3 Quality management systems;
4 Promoting & motivating experimentation;	2 Additional innovation: "improvement of a previously defined design: The improvement is observed in individual elements, but the basic concepts of the design and realization of the production of the product and service, as well as the connections between them do not change."	4 Business management, and "total characteristics of products and services are: • Based on strategy—TQM is considered an important part of the business' vision, mission, objectives, action plan, etc. • Client in focus and center of attention; • Attempt to realize the connection between clients' requirements with product/service features; • Scientific problem-solving framework—problem-solving tools and techniques; • Long-term commitment—quality is a journey, not a destination; • Teamwork—setting quality improvement objectives, and processes; • Continuous improvement process—a quality achievement never ends; • Education and training—of employees to achieve total quality; • Freedom through control—to express views, perform quality improvement actions; • Unity of purpose—based on the objectives of the organization; • Employee involvement and empowerment—through competencies, and quality management.
5 Constructing promotes the environment, specific promotion inside organizations;		
6 Promoting risk-taking initiatives;		
7 Mastering auto-competition;	3 Architectural innovation: "innovation that changes only the relationships between them [the core design concepts]"	
8 Promoting supportive values and ideas on creative ideas;		
9 Offering chances to choose and explore;		
10 Offering conditions for the development of auto-management;	4 Modular Innovation: "innovation that changes only the core design concepts of a technology" [66].	
11 Learning approaches, helping the performance of creativity;		
12 Supporting and establishing a balanced environment" [67]		

Table 1.
Relations between innovation, creativity, and quality management [64, 65].

3. Research framework, the purpose of the case study

The amount of innovation, originality, and ISO certifications in the form granted for each nation for which data could be gathered served as the framework for the study of the global entrepreneurship ecosystem [68].

The purpose of this research is to fill this gap in the literature regarding the relationships between creativity and innovation, creativity and quality management, and quality management and innovation. Additionally, this research has adopted a theory-building approach with the following objectives [68]:

1. RQ1: Is there any relationship between innovation and creativity?

2. RQ2: Is there any relationship between quality management and innovation?

3. RQ3: Is there any relationship between quality management and creativity.

4. Methodology

Despite acknowledging the value of quality management, creativity, and innovation in business, previous theoretical and empirical research has not been able to scientifically explain how these factors interact and influence one another. Therefore, the purpose of this study is to develop a theory that, supported by data, facts, conclusions, and analysis, confirms the relationships among quality management, creativity, and innovation.

In contrast, this study considers (1) labor as non-homogeneous, as it is in reality, and (2) land as it is in reality, not homogenous, in opposition to practically all traditional studies connected to the components of production (as each country has specific topsoil, soil, and subsoil, which affect production, capital flow, and human resource management). (3) Economic agents, people, and families are taken into account as they are, not as consumers of the same or comparable goods and services in the same quality and quantity; (4) changes in income and prices are viewed as ordinary occurrences rather than as solid; (5) shifts in public and private sector management as a result of managers', owners', and employees' varying levels of experience; (6) knowledge, skills, and competencies are not distributed equally; and (7) capital is not a requirement for job simulation. These qualities are precisely what bring this art so closely to life.

4.1 Selection of case

Three key factors have been considered in this study: theoretical elements, appropriate existing data that might be further processed by the study's author, and actual examples of the connections among quality management, creativity, and innovation. The study questions were created in light of this. These are the research queries:

- RQ1: Is there any relationship between innovation and creativity?

- RQ2: Is there any relationship between innovation and quality management?

- RQ3: Is there any relationship between creativity and quality management?

Based on these research questions, three pairs of hypotheses were constructed:
Relationships between innovation and creativity (RQ1).
H01 – There is no relationship between innovation and creativity.
H11 – There is a strong relationship between innovation and creativity.

Relationships between innovation and quality management (RQ2).
H02 – There is no relationship between innovation and quality management.
H12 – There is a strong relationship between innovation and quality management.
Relationships between creativity and quality management (RQ3).
H03 – There is no relationship between creativity and quality management.
H13 – There is a strong relationship between creativity and quality management.

4.2 Collection of data

- Data about creativity taken from The 2020 Creativity Index Report (**Table 2**).
- Data about innovation taken from the Global Innovation Index Report 2020 (**Table 2**).
- Data on the application of the ISO standards—ISO Report 2015-2020 (**Table 2**).
- The ISO standard index for every country was calculated by dividing the total number of business entities in the country by the number of emissions certified with ISO standards for each country, as given in a table (**Table 2**).

Only the following two forms of data were used in the development of this study: (1) indexes and international websites and (2) data that the author processed on the ISO Certificates Index calculated for each country [64, 65].

No	Countries	Creative output index	GII	Creative output inverse index	Innovation inverse index
1	Switzerland	2	1	131	132
2	Luxembourg	3	2	130	110
3	United Kingdom	4	3	129	129
4	Sweden	5	4	128	131
5	France	6	6	127	122
6	Netherlands	7	7	126	127
7	Malta	9	8	124	106
8	Iceland	10	9	123	116
9	Germany	11	10	122	123
10	United States of America	12	11	121	130
11	Denmark	13	12	120	124
12	China	14	13	119	121
13	Estonia	15	15	118	112
14	Finland	16	16	117	126
15	Singapore	17	17	116	125
16	Japan	18	18	115	120
17	Canada	19	19	114	117
18	Cyprus	20	20	113	105

No	Countries	Creative output index	GII	Creative output inverse index	Innovation inverse index
19	Bulgaria	21	21	112	98
20	Czech Republic	22	22	111	109
21	New Zealand	23	23	110	107
22	Australia	24	24	109	108
23	Norway	25	25	108	113
24	Portugal	26	26	107	102
25	Austria	27	27	106	115
26	Ireland	29	28	104	114
27	Israel	30	29	103	118
28	Mauritius	31	30	102	81
29	Spain	32	31	101	103
30	Montenegro	33	32	100	83
31	Italy	34	33	99	104
32	Turkey	35	34	98	92
33	Belgium	36	35	97	111
34	Malaysia	37	36	96	97
35	Slovenia	38	37	95	101
36	Latvia	39	38	94	95
37	United Arab Emirates	40	39	93	100
38	Lithuania	41	40	92	94
39	Viet Nam	42	41	91	44
40	Slovakia	43	42	90	96
41	United Rep. Tanzania	44	43	89	42
42	Costa Rica	45	44	88	77
43	Iran (Islamic Republic of)	46	45	87	73
44	Hungary	47	46	86	99
45	Ukraine	48	47	85	84
46	Armenia	49	48	84	64
47	Poland	50	49	83	93
48	Jamaica	51	50	82	59
49	Mexico	52	51	81	78
50	Croatia	54	52	79	91
51	Thailand	55	53	78	90
52	Russian Federation	56	54	77	88
53	El Salvador	57	55	76	37
54	Panama	58	56	75	50
55	Chile	60	57	73	80

No	Countries	Creative output index	GII	Creative output inverse index	Innovation inverse index
56	Qatar	63	59	70	65
57	Uruguay	64	60	69	68
58	Philippines	65	61	68	82
59	Brazil	66	62	67	76
60	Azerbaijan	67	63	66	53
61	India	68	65	65	87
62	Greece	69	66	64	86
63	Morocco	70	67	63	56
64	Romania	72	68	61	85
65	Argentina	73	69	60	60
66	Georgia	74	70	59	70
67	Guatemala	75	71	58	32
68	Serbia	76	72	57	79
69	Peru	77	73	56	63
70	Saudi Arabia	78	74	55	67
71	South Africa	79	75	54	72
72	Tunisia	80	77	53	62
73	Albania	81	78	52	49
74	Colombia	82	79	51	66
75	North Macedonia	83	80	50	74
76	Dominican Republic	84	81	49	40
77	Brunei Darussalam	85	82	48	51
78	Pakistan	87	83	46	34
79	Jordan	88	84	45	52
80	Kuwait	89	85	44	61
81	Lao People's Democr, Rep	90	86	43	16
82	Indonesia	91	87	42	46
83	Lebanon	92	90	41	42
84	Belarus	93	92	40	71
85	Ghana	94	93	39	21
86	Kenya	95	94	38	48
87	Guinea	96	95	37	3
88	Malawi	97	96	36	26
89	Cambodia	98	97	35	24
90	Bosnia and Herzegovina	99	98	34	58
91	Sri Lanka	100	99	33	38
92	Honduras	102	100	31	25

No	Countries	Creative output index	GII	Creative output inverse index	Innovation inverse index
93	Trinidad and Tobago	103	101	30	36
94	Egypt	104	102	29	39
95	Namibia	105	103	28	33
96	Bahrain	106	105	27	55
97	Tajikistan	107	106	26	30
98	Nepal	108	107	25	22
99	Senegal	109	108	24	28
100	Kazakhstan	110	109	23	54
101	Botswana	112	111	21	27
102	Uzbekistan	113	112	20	47
103	Yemen	114	115	19	2
104	Mozambique	115	116	18	11
105	Nigeria	116	117	17	15
106	Rwanda	117	118	16	31
107	Algeria	118	119	15	12
108	Togo	119	120	14	8
109	Kyrgyzstan	120	121	13	35
110	Bangladesh	123	122	10	27
111	Cameroon	124	123	9	10
112	Zambia	125	125	8	12
113	Uganda	126	126	7	14
114	Ethiopia	127	128	6	7
115	Benin	128	129	5	5
116	Burkina Faso	129	130	4	18
117	Angola	130	131	3	1
118	Niger	132	132	2	4

Table 2.
Innovation index and creativity index [35].

4.3 Analysis of data

1. The number of businesses in each nation that are ISO-certified was gathered from the ISO website.

2. The total number of business entities was gathered from numerous websites and other written materials including houses of society, yearly reports of the nation, institutes of statistics, and so on.

3. To create the ISO index, the total number of businesses in each country was divided by the number of companies that were ISO-certified (**Table 2**).

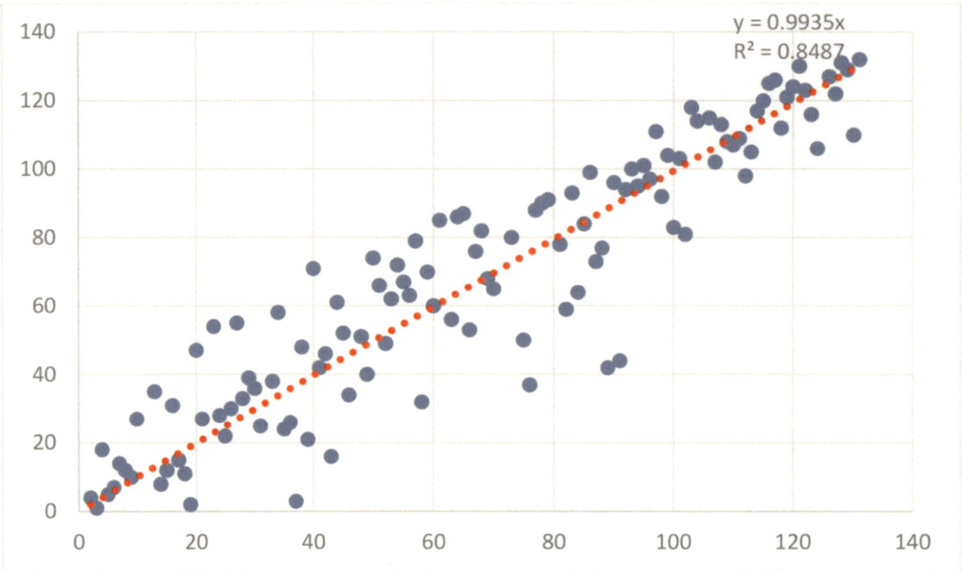

Figure 2.
Regression line innovation and creativity.

4. The creative index (**Table 2**) was used to gather the information for the World countries' creative rankings [69].

5. Data on the Global Innovation Index (**Table 2**) were taken from rankings of World nations' innovation performance (World Intellectual Property Organization, 13th Edition).

6. A regression analysis was done between the Innovation Index and the Creativity Indices (**Table 2**, together with the respective **Figure 2** and regression tables), Innovation Index and ISO Standards Index (**Table 3**, together with the respective **Figure 3** and regression tables), and among Creativity Index and ISO Standards Index (**Table 4**, together with the respective **Figure 4** and regression tables).

No	Countries	Innovation inverse index	ISO standard Index
1	Switzerland	132	0.022
2	Luxembourg	110	0.00231
3	United Kingdom	129	0.00884
4	Sweden	131	0.00575
5	France	122	0.0054
6	Netherlands	127	0.0072
7	Malta	106	0.00596
8	Iceland	116	0.00203
9	Germany	123	0.021
10	United States of America	130	0.00095

No	Countries	Innovation inverse index	ISO standard Index
11	Denmark	124	0.0071
12	China	121	0.00434
13	Estonia	112	0.0071
14	Finland	126	0.0082
15	Singapore	125	0.0153
16	Japan	120	0.01123
17	Canada	117	0.0052
18	Cyprus	105	0.0053
19	Bulgaria	98	0.0129
20	Czech Republic	109	0.0207
21	New Zealand	107	0.00321
22	Australia	108	0.00576
23	Norway	113	0.00695
24	Portugal	102	0.0114
25	Austria	115	0.0173
26	Ireland	114	0.0136
27	Israel	118	0.018
28	Mauritius	81	0.00203
29	Spain	103	0.0146
30	Montenegro	83	0.0146
31	Italy	104	0.021
32	Turkey	92	0.00132
33	Belgium	111	0.00467
34	Malaysia	97	0.0115
35	Slovenia	101	0.0127
36	Latvia	95	0.01173
37	United Arab Emirates	100	0.01185
38	Lithuania	94	0.0099
39	Viet Nam	44	0.0131
40	Slovakia	96	0.0166
41	United Republic of Tanzania	42	0.000053
42	Costa Rica	77	0.00428
43	Iran (Islamic Republic of)	73	0.0288
44	Hungary	99	0.00925
45	Ukraine	84	0.00121
46	Armenia	64	0.00124
47	Poland	93	0.00656
48	Jamaica	59	0.00406

No	Countries	Innovation inverse index	ISO standard Index
49	Mexico	78	0.00209
50	Croatia	91	0.0149
51	Thailand	90	0.00505
52	Russian Federation	88	0.0019
53	El Salvador	37	0.00147
54	Panama	50	0.00536
55	Chile	80	0.0293
56	Uruguay	68	0.0088
57	Philippines	82	0.00544
58	Brazil	76	0.00343
59	India	87	0.00082
60	Morocco	56	0.01886
61	Romania	85	0.0144
62	Argentina	60	0.00951
63	Georgia	70	0.01136
64	Guatemala	32	0.0164
65	Serbia	79	0.0189
66	Peru	63	0.00162
67	Saudi Arabia	67	0.00343
68	South Africa	72	0.00196
69	Tunisia	62	0.00212
70	Albania	49	0.0043
71	Colombia	66	0.00558
72	North Macedonia	74	0.0191
73	Dominican Republic	40	0.00311
74	Brunei Darussalam	51	0.0169
75	Pakistan	34	0.022
76	Jordan	52	0.00282
77	Lao People's Democr. Rep	16	0.00054
78	Indonesia	46	0.000018
79	Lebanon	42	0.00354
80	Ghana	21	0.00317
81	Kenya	48	0.00022
82	Guinea	3	0.00595
83	Malawi	26	0.00215
84	Cambodia	24	0.00028
85	Bosnia and Herzegovina	58	0.0195
86	Sri Lanka	38	0.0019

No	Countries	Innovation inverse index	ISO standard Index
87	Honduras	25	0.00143
88	Trinidad and Tobago	36	0.0048
89	Egypt	39	0.00094
90	Namibia	33	0.00125
91	Bahrain	55	0.0093
92	Tajikistan	30	0.008
93	Nepal	22	0.00016
94	Senegal	28	0.00046
95	Kazakhstan	54	0.002
96	Botswana	27	0.00106
97	Uzbekistan	47	0.00075
98	Yemen	2	0.025
99	Mozambique	11	0.01276
100	Nigeria	15	0.000014
101	Rwanda	31	0.00039
102	Algeria	12	0.00034
103	Togo	8	0.00036
104	Kyrgyzstan	35	0.01179
105	Bangladesh	27	0.00126
106	Cameroon	10	0.00046
107	Zambia	12	0.00004
108	Uganda	14	0.00014
109	Ethiopia	7	0.00113
110	Benin	5	0.01044
111	Burkina Faso	18	0.00044
112	Angola	1	0.00251
113	Niger	4	0.00698

Table 3.
Innovation index [35] and ISO standard index.

Relations between innovation and creativity. (118 Countries)
Relations between innovation and creativity (RQ1).
H01 – There is no relationship between innovation and creativity.
H11 – There is a strong relationship between innovation and creativity.

Regression Statistics	
Multiple R	0.982182
R Square	0.964681

Adjusted R Square	0.956134
Standard Error	14.551
Observations	118

ANOVA					
	df	*SS*	*MS*	*F*	*Significance F*
Regression	1	676615.4	676615.4	3195.629	2.83E-86
Residual	117	24772.59	211.7315		
Total	118	701388			

	Coefficients	**Standard Error**	**t Stat**	**P-value**	**Lower 95%**	**Upper 95%**	**Lower 95.0%**	**Upper 95.0%**
Intercept (b)	0	#N/A	#N/A	#N/A	#N/A	#N/A	#N/A	#N/A
Innv. INDEX (a)	0.970979	0.017176	56.52989	8.58E-87	0.936963	1.004996	0.936963	1.004996

	y= ax+b	
Model	Y = 0.9935x	y = 0.9935x
	R^2 = 0.9647	

The results show that Innovation at a level of 96.47% has the explanation of indication of the Creativity factor [2].

Relations between the Innovation index and the Creativity index are strong (r = 0.982182).

Regression equation y = 0.9935x

R^2 coefficient = 0.09647

Correlation coefficient "r" = 0.982182.

Hypothesis:

H01 – There is no relationship between innovation and creativity.

– the model is not good, with the security level α=0.05.

H11 – There is a strong relationship between innovation and creativity.

– the model is good.

By ANOVA $F_{llog} > F_{crit}$, F Significance F (probability getting these results) $< \alpha$= 0.05.

H0 is not valuable, H1 has been verified, with a significance level of 0.05 or a level of reliability = 95 %.

Coefficients are the values of the correlation coefficient.

Relations between innovation and quality management (RQ2).

H02 – There is no relationship between innovation and creativity.

H12 – There is a strong relationship between innovation and creativity.

Regression Statistics	
Multiple R	0.732825
R Square	0.537033
Adjusted R Square	0.528024
Standard Error	0.006937
Observations	112

ANOVA					
	Df	*SS*	*MS*	*F*	*Significance F*
Regression	1	0.006196	0.006196	128.7578	3.16E-20
Residual	111	0.005341	4.81E-05		
Total	112	0.011537			

	Coefficients	*Standard Error*	*t Stat*	*P-value*	*Lower 95%*	*Upper 95%*	*Lower 95.0%*	*Upper 95.0%*
Intercept (b)	0	#N/A	#N/A	#N/A	#N/A	#N/A	#N/A	#N/A
Innov INDEX (a)	9.57E-05	8.43E-06	11.34715	2.81E-20	7.9E-05	0.000112	7.9E-05	0.000112

	y= ax+b	
Model	Y = 1E-0.4x	y = 0.000464 (Innovation Index)x – 0.01104 (intercept)
	R^2 = 0.54	

At ISO 54.0% can be explained under the indication of the Innovation factor
The relation (Connection) between them is strong (r = 0.732825).
Regression equation y = 1E-0.4x
R^2 coefficient = 0.54
Correlation coefficient "r" = 0.732825.
Hypothesis:
H02 – There is no relationship between innovation and quality management.
– the model is not good, with the security level α=0.05.
H12 – There is a strong relationship between innovation and quality management.
– the model is good.
By ANOVA F_{llog}> F_{crit} F Significance F (probability getting these results) < α= 0.05.

H0 is not valuable and H1 has been verified, with a significance level of 0.05 or a level of reliability = 95 %.

Coefficients are the values of the correlation coefficient.

Relations between innovation and ISO certificates (113 Countries)

Relations between creativity and quality management (RQ3).

H03 – There is no relationship between innovation and creativity.

H13 – There is a strong relationship between innovation and creativity.

Regression Statistics	
Multiple R	0.725443
R Square	0.526268
Adjusted R Square	0.517259
Standard Error	0.007017
Observations	112

ANOVA					
	df	*SS*	*MS*	*F*	*Significance F*
Regression	1	0.006071	0.006071	123.3098	1.14E-19
Residual	111	0.005465	4.92E-05		
Total	112	0.011537			

	Coefficients	*Standard Error*	*t Stat*	*P-value*	*Lower 95%*	*Upper 95%*	*Lower 95.0%*	*Upper 95.0%*
Intercept (b)	0	#N/A	#N/A	#N/A	#N/A	#N/A	#N/A	#N/A
Creativity INDEX (a)	9.54E-05	8.59E-06	11.1045	1.01E-19	7.84E-05	0.000112	7.84E-05	0.000112

	y= ax+b	
Model	Y = 1E-0.4x	y = 1E-0.4x
	R^2 = 0.538	

ISO 58.0% under Creativity factor indication

The relation is strong (r = 0.725443).

Hypothesis:

H03 – There is no relationship between creativity and quality management.

– the model is not good, with the security level α=0.05.

H13 – There is a strong relationship between creativity and quality management.

– the model is good.

By ANOVA $F_{llog} > F_{crit}$ F Significance F (probability getting these results) < α= 0.05

H0 is not valuable and H1 has been verified, with a significance level of 0.05 or a level of reliability = 95 %.

Coefficients are the values of the correlation coefficient.

Relations between creativity and ISO certificates. (113 Countries)

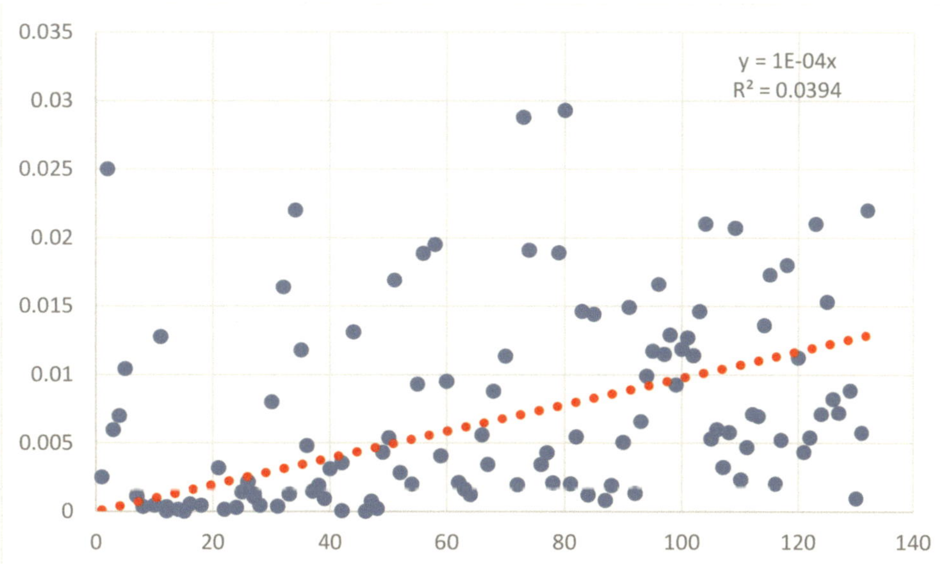

Figure 3.
Regression line innovation inverse IndeazSx and ISO standard index.

No	Countries	Creative output inverse index	ISO standard index
1	Switzerland	131	0.022
2	Luxembourg	130	0.00231
3	United Kingdom	129	0.00884
4	Sweden	128	0.00575
5	France	127	0.0054
6	Netherlands	126	0.0072
7	Malta	124	0.00596
8	Iceland	123	0.00203
9	Germany	122	0.021
10	United States of America	121	0.00095
11	Denmark	120	0.0071
12	China	119	0.00434
13	Estonia	118	0.0071
14	Finland	117	0.0082
15	Singapore	116	0.0153
16	Japan	115	0.01123
17	Canada	114	0.0052
18	Cyprus	113	0.0053
19	Bulgaria	112	0.0129
20	Czech Republic	111	0.0207

No	Countries	Creative output inverse index	ISO standard index
21	New Zealand	110	0.00321
22	Australia	109	0.00576
23	Norway	108	0.00695
24	Portugal	107	0.0114
25	Austria	106	0.0173
26	Ireland	104	0.0136
27	Israel	103	0.018
28	Mauritius	102	0.00203
29	Spain	101	0.0146
30	Montenegro	100	0.0146
31	Italy	99	0.021
32	Turkey	98	0.00132
33	Belgium	97	0.00467
34	Malaysia	96	0.0115
35	Slovenia	95	0.0127
36	Latvia	94	0.01173
37	United Arab Emirates	93	0.01185
38	Lithuania	92	0.0099
39	Viet Nam	91	0.0131
40	Slovakia	90	0.0166
41	United Republic of Tanzania	89	0.000053
42	Costa Rica	88	0.00428
43	Iran (Islamic Republic of)	87	0.0288
44	Hungary	86	0.00925
45	Ukraine	85	0.00121
46	Armenia	84	0.00124
47	Poland	83	0.00656
48	Jamaica	82	0.00406
49	Mexico	81	0.00209
50	Croatia	79	0.0149
51	Thailand	78	0.00505
52	Russian Federation	77	0.0019
53	El Salvador	76	0.00147
54	Panama	75	0.00536
55	Chile	73	0.0293
56	Uruguay	69	0.0088
57	Philippines	68	0.00544
58	Brazil	67	0.00343

No	Countries	Creative output inverse index	ISO standard index
59	India	65	0.00082
60	Morocco	63	0.01886
61	Romania	61	0.0144
62	Argentina	60	0.00951
63	Georgia	59	0.01136
64	Guatemala	58	0.0164
65	Serbia	57	0.0189
66	Peru	56	0.00162
67	Saudi Arabia	55	0.00343
68	South Africa	54	0.00196
69	Tunisia	53	0.00212
70	Albania	52	0.0043
71	Colombia	51	0.00558
72	North Macedonia	50	0.0191
73	Dominican Republic	49	0.00311
74	Brunei Darussalam	48	0.0169
75	Pakistan	46	0.022
76	Jordan	45	0.00282
77	Lao People's Democr. Rep.	43	0.00054
78	Indonesia	42	0.000018
79	Lebanon	41	0.00354
80	Ghana	39	0.00317
81	Kenya	38	0.00022
82	Guinea	37	0.00595
83	Malawi	36	0.00215
84	Cambodia	35	0.00028
85	Bosnia and Herzegovina	34	0.0195
86	Sri Lanka	33	0.0019
87	Honduras	31	0.00143
88	Trinidad and Tobago	30	0.0048
89	Egypt	29	0.00094
90	Namibia	28	0.00125
91	Bahrain	27	0.0093
92	Tajikistan	26	0.008
93	Nepal	25	0.00016
94	Senegal	24	0.00046
95	Kazakhstan	23	0.002
96	Botswana	21	0.00106

No	Countries	Creative output inverse index	ISO standard index
97	Uzbekistan	20	0.00075
98	Yemen	19	0.025
99	Mozambique	18	0.01276
100	Nigeria	17	0.000014
101	Rwanda	16	0.00039
102	Algeria	15	0.00034
103	Togo	14	0.00036
104	Kyrgyzstan	13	0.01179
105	Bangladesh	10	0.00126
106	Cameroon	9	0.00046
107	Zambia	8	0.00004
108	Uganda	7	0.00014
109	Ethiopia	6	0.00113
110	Benin	5	0.01044
111	Burkina Faso	4	0.00044
112	Angola	3	0.00251
113	Niger	2	0.00698

Table 4.
Creativity index [35] and ISO standard index.

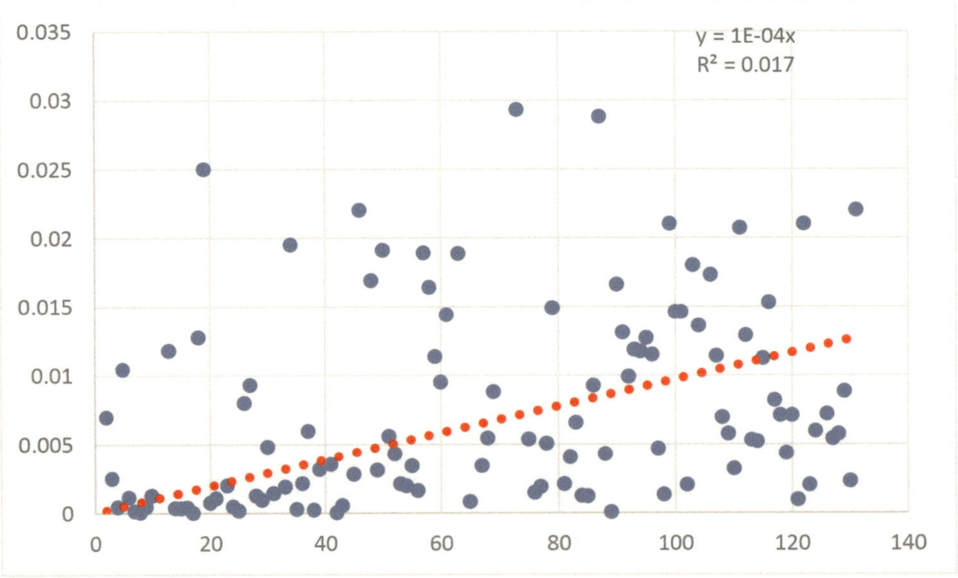

Figure 4.
Regression line creativity and ISO standard.

5. Summary output

The relationship among quality management, creativity, and innovation has been the subject of three significant conclusions from this study. This is not just due to the theoretical approach, but also because this study's conclusions are the first to be statistically confirmed and show a strong link among quality management, creativity, and innovation.

In this study, an ISO certificate index is used to rank nations based on the number of ISO certificates issued in a country divided by the number of business organizations already operating there during the same period [64, 65].

6. Research's context

The idea that land, labor, and capital are the most important variables in both the management and economic aspects of an enterprise appears to be outdated in the twenty-first century. Considering only entrepreneurial ability, the fourth factor of production is not a viable option because it cannot include the elements of innovation and creativity, which are the traits shared by both employees and employers as well as business owners and employers.

Quality, creativity, and innovation may and should be the foundation for gaining, retaining, and managing economic advantages throughout the short, medium, and long term, as evidenced by events over the past at least 45 to 50 years in a world that is becoming more international. The study's findings demonstrate unequivocally that nations with high ISO standards also tend to have high levels of both creativity and innovation indices.

Although previous studies have demonstrated the value of quality management, creativity, and innovation, few studies address the psychological aspects of this relationship. Additionally, there are no statistically supported explanations for how quality management, creativity, and innovation interact to produce economic benefits.

7. Discussion

The level of innovation, creativity, and ISO standard certificates in the worldwide entrepreneurial ecosystem served as the research's framework.

The main findings of a regression analysis of the relationships between innovation and creativity, innovation and ISO 9001 standards, and creativity and ISO 9001 standards are that there is a significant relationship among innovation and creativity, innovation and ISO 9001 standards, and creativity and ISO 9001 standards as well.

- Theoretical and practical consideration

In terms of the theory, the final findings suggest a fresh direction for further investigation into the connections among quality management, creativity, and innovation, as well as among other economic indicators, and so on, which is still regarded as uncharted territory.

According to the research, it is crucial to treat quality management, creativity, and innovation as a triple platform that paves the way to better economic advantage for both individual firms and a nation's economy.

8. Limitations and avenues for further research

This essay aims to draw attention to current issues by recognizing the connections between creativity, innovation, and quality control as well as the options for dealing with them. This paper's research opens a window for academics and other practitioners in these fields as we are now much closer to being able to design studies that will provide better answers to such questions. While questions of the processes that facilitate these relationships are the subject of further investigation, there is currently enough information available to provide some definitive answers to the questions of these relationships.

This study, one of the first to highlight the connections between quality management, creativity, and innovation, first and the psychological aspect of these relationships, was conducted using extensive data on ISO certificates issued for each country as well as data on quality management, creativity, and innovation for 2020. Reliable data on some business entities registered in the countries are also taken into consideration. To conduct a comprehensive investigation of these correlations, however, more data are needed for different periods of combination research involving psychological techniques that might be started from the perspective of creativity.

Studies that take into account both the quality and amount of investments can be started from the perspective of innovation.

Combination studies on the industries most impacted by the interaction among creativity, innovation, and quality management can be started from the standpoint of quality management.

The paper suggests that the following studies should:

1. Within the design, ask the right process questions and gather information to show how creativity, innovation, and quality control are related globally.

2. Information from correlational studies should be sufficiently detailed to provide information on current organizational business aspects, such as human resources, strategic management, organizational behavior, supply chain management, marketing strategies, the digital economy, and areas that call for a complete eclipse of the required changes.

3. Considering that researchers should not take into account every possible interaction but instead strive for data collection efforts to evaluate the interactions of specifically identified variables, conceptualize potential interactions between variables that may be particularly pertinent to the topics under study [70].

9. Conclusions and recommendations

1. This study reveals a substantial correlation between quality management and creativity, between quality management and innovation, and between creativity

and innovation, drawing three significant conclusions for the first time. These findings are not merely theoretical assertions; they have been statistically supported by the use of regression analysis for the three examples, in a comprehensive manner that takes into account the psychological components of the problem.

2. The ISO index used in this study ranks nations according to the ratio of business entities present in each nation over the same period to the number of ISO certifications issued in that nation [64, 65].

3. Strong statistical links between innovation and creativity, between creativity and ISO certification, and between innovation and ISO certification suggest that, in addition to the effective and efficient use of other production factors (labor, materials, etc.), the promotion of creativity, innovation, and quality management as a non-fixed factor is necessary to achieve a competitive and comparative advantage.

4. The relationship between the cultures of innovation, creativity, and quality management is strong at the global level, and this mostly pertains to the cultures of entrepreneurship and conducting business in reaction to the crisis and the immediate aftermath of the disaster.

5. This study clarifies and confirms that natural resources do not constitute any barrier regarding productivity and levels of economic growth, despite the belief of some experts that the world is headed toward poverty because natural resources are unchangeable (fixed).

6. There are at least two ways to increase the productivity of the global economy while avoiding the constraints of fixed factors of production (land, labor, and capital), first by using quality management, creativity, and innovation, and second by combining these three approaches both individually and with other entrepreneurship-related components.

7. There is a propensity to innovate and develop to reduce the use of fixed factors of production, which gives rise to the idea that technical advancements enable fixed forces of production to no longer serve as constraints on productivity growth and global economic expansion. The connections between quality management, creativity, and innovation can also be explained using the same logic and justification.

8. The factors of production and the cycles of their usage are tied to the history of economic philosophy, economic systems, and the growth of the global economy. One of these cycles involves the use of fixed factors of production like land, labor, and capital, while the other involves the use of quality control, innovation, and creativity (skills, knowledge, and entrepreneurial competencies—non-fixed factors of production). The cycles of fixed and limited factors of production (land, labor, and capital) have dominated at certain points in world history and economic development, whereas the cycles of non-fixed factors of production (innovation, creativity, focus on quality, and entrepreneurial spirit) have dominated at other points in world history and economic development.

Investigated literature

1. Agius C, Devine K. Neutrality: A dead concept? A reprise. Cooperation and Conflict. 2011;**46**:265-284. DOI: 10.1177/0010836711416955

2. Mohammed A, Nordin N. A theoretical evidence: The mediating role of creativity on the relationship between knowledge management and entrepreneurial orientation toward innovation performance In the Jordanian SMEs. International Journal of Scientific & Technology Research. 2020;**9**(3): 3339-3355. Available from: https://ijstr.org/paper-references.php?ref=IJSTR-0320-31580

3. Conn A-M, Szilagyi MA, Alpert-Gillis L, Webster-Stratton C, Manly JT, Goldstein N, et al. Pilot randomized controlled trial of foster parent training: A mixed-methods evaluation of parent and child outcomes. Children and Youth Se rvices Review. 2018;89:188-197. doi: 10.1016/j.childyouth.2018.04.035. Available from: https://www.sciencedirect.com/science/article/abs/pii/S0190740917307429?via%3Dihub

4. Kleinfeld A, Martens A, Bribery. In: Idowu S, Schmidpeter R, Capaldi N, Zu L, Del Baldo M, Abreu R. (Hrsg.): Encyclopedia of Sustainable Management. Cham: Springer; 2020. Available from: https://www.htwg-konstanz.de/en/resea rch-and-transfer/institutes-and-laboratories/kicg/publications/professors-me mbers/annette-kleinfeld/

5. Appelbaum SH, Steed AJ. The critical success factors in the client-consulting relationship. Journal of Management Development. 2005;**24**:68-93. DOI: 10.1108/02621710510572362

6. Astor H. Mediator neutrality: Making sense of theory and practice. Social & Legal Studies. 2007;**16**:221-239. DOI: 10.1177/0964663907076531

7. Averill JR. Individual differences in emotional creativity: Structure and correlates. Journal of Personality. 1999;**67**(2):331-371

8. Azadegan A, Kolfschoten G. An assessment framework for practicing facilitators. Group Decision and Negotiation. 2014;**23**:1013-1045. DOI: 10.1007/s10726-012-9332-4

9. Barczak G, Lassk F, Mulki J. Antecedents of team creativity: An examination of team emotional intelligence, team trust, and collaborative culture. Creativity and Innovation Management. 2010;**19**:332-345. DOI: 10.1111/j.1467-8691.2010.00574.x

10. Bens I. Facilitating with Ease! Core Skills for Facilitators, Team Leaders, Members, Managers, Consultants, and Trainers. New York: John Wiley & Sons; 2012

11. Berman EM, Kim C-G. Creativity management in public organizations. Public Performance & Management Review. 2010;**33**:619-652. DOI: 10.2753/pmr1530-9576330405

12. Bernard SE, Folger JP, Weingarten HR, Zumeta ZR. The neutral mediator: Value dilemmas in divorce mediation. Conflict Resolution Quarterly. 1984;**1984** (4):61-74. DOI: 10.1002/crq.39019840406

13. Berry M. Changing perspectives on facilitation skills development. Journal of European Industrial Training. 1993;**17**(3):23. DOI: 10.1108/03090599310 026355

14. Braun V, Clarke V. Using thematic analysis in psychology. Qualitative Research in Psychology. 2006;**3**(2):77-101. DOI: 10.1191/1478088706qp063oa

15. Greenhalgh C, Rogers M. Innovation, Intellectual Property, and Economic Growth. Princeton, New Jersey: Published by Princeton University Press; 2010 ISBN 978-0-691-13798-8. doi:10.1515/9781400832231. Available from: https://www.jstor.org/stable/j.ctt1zgwjjb. Available from: https://www.degruyter.com/document/doi/10.1515/9781400832231/html

16. Cydney Spohn. 2008. Teacher perspectives on No child left behind and arts education: A case study. Arts Education Policy Review. 109(4); 3-12doi: 10.3200/AEPR.109.4.3-12. Available from: https://www.tandfonline.com/doi/abs/10.3200/AEPR.109.4.3-12. Published online: 07 Aug 2010

17. Doloreux, D. 2004. Regional networks of small and medium sized enterprises: Evidence from the metropolitan area of Ottawa in Canada. European Planning Studies, Vol. 12, No. 2, 173-189. ISSN 0965-4313

18. Engelberger JF. Robotics in Practice: Future Capabilities. Electronic Servicing & Technology Magazine; 1982. DOI: 10.1145/2185376.2185393

19. Ernst, H. 2001. Patent applications and subsequent changes of performance: Evidence from time-series cross-section analyses on the firm level. Research Policy, Vol 30, No. 1, 143-157. ISSN 0048-7333

20. Hermundsdottir F, Aspelund A. Sustainability innovations and firm competitiveness: A review. Journal of Cleaner Production. 2021;**280**(1):124715

21. O'Regan G. A Practical Approach to Software Quality. Springer Science and Business Media LLC; 2002. Available from: https://www.scirp.org/%28S%28lz5mqp453ed snp55rrgjct55%29%29/reference/referencespapers.aspx?referenceid=1131855

22. Hicks JR. The Theory of Wages. London: Macmillan; 1932

23. Gupta H, Barua MK. Modelling cause and effect relationship among enablers of innovation in SMEs. Benchmarking: An International Journal. 2018;**25**(5):1597-1622. DOI: 10.1108/BIJ-03- 2017-0050

24. Hjorth D. Creating space for play/invention – Concepts of space and organizational entrepreneurship. Entrepreneurship and Regional Development. 2004;**16**(5):413-432

25. Hoegl M, Gibbert M, Mazursky D. Financial constraints in innovation projects: When is less more? Research Policy. 2008;**37**(8):1382-1139

26. Harvey L, Williams J. Fifteen Years of Quality in Higher Education, Quality in Higher Education. 2010;**16**(1):3-36. DOI: 10.1080/13538321003679457

27. Kim J, Lee J, Choi D. Designing emotionally evocative homepages: An empirical study of the quantitative relations between design factors and emotional dimensions. International Journal of Human-Computer Studies. 2003;**59** (6):899-940. DOI: 10.1016/j.ijhcs.2003.06.002. Available from: https://www.sciencedirect.com/science/article/abs/pii/S1071581903001320?via%3Dihub

28. Jong JPJ. Individual Innovation: The Connection between Leadership and Employees' Innovative Work Behavior. The University of Amsterdam. UvA-DARE (Digital Academic Repository); 2007. ISBN 978-90-371-0725-8. Available from: https://ideas.repec.org/p/eim/papers/r200604.html

29. Wilde J. How substitutable are fixed factors in production? evidence from pre-industrial England. Germany: MPRA Paper, University Library of Munich; 2012. Available from: https://EconPapers.repec.org/RePEc:pra:mprapa:39278, https://econpapers.repec.org/paper/ pramprapa/39278.htm

30. Kiran DR. Total Quality Management: Key Concepts and Case Studies. Amsterdam: Elsevier Inc; 2017

31. Caswell JA, Bredahl ME, Hooker NH. How quality management metasystems are affecting the food industry. Review of Agricultural Economics, Agricultural and Applied Economics Association. 1998;**20**(2):547-557

32. Lijster T, editor. The Future of the New: Artistic Innovation in Times of Social Acceleration. Valiz: Arts in Society; 2018. ISBN 9789492095589. Available from: http://www.westdenhaag.nl/information/exhibitions/18_10_The_Future_of_the_New/more1/TheFutureOfTheNew.pdf [Retrieved 10 Sep 2020]

33. Loo S. Creative working in knowledge economy. In: Series: Routledge Advances in Organizational Learning and Knowledge Management. New York: Routledge; 2017

34. Dejan M, Vlade M, Milan G. Da li najbolja preduzeća u Srbiji kreiraju vrednost? Ekonomika preduzeća. 2015;**63**(7–8):337-353 ISSN: 0353-443X, e- ISSN: 2406-1239. Available from: http://scindeks.ceon.rs/Article.aspx?artid=0353-443X1508337M

35. Agovino M, Corbisiero F. Diversity management policies for the inclusion of LGBT people into the labour market: The case of Naples. Papers in Applied Geography. 2020;**7**(1):50-66. DOI: 10.1080/23754931.2020.1815236. Available from: https://www.tandfonline.com/doi/full/10.1080/23754931. 2020.1815236. Published online: 01 Sep 2020

36. Gibbert M, Scranton P. Constraints as sources of radical innovation? Insights from jet propulsion development. Management & Organizational History. 2009;**4**(4):385-399. DOI: 10.1177/1744935909341781

37. Lewis M, Moultrie J. The Organizational Innovation. Laboratory. 2005. DOI: 10.1111/j.1467-8691.2005 .00327.x

38. Anderson NR, Potočnik K, Zhou J. Innovation and creativity in organizations: A state-of-the-science review, prospective commentary, and guiding framework. Journal of Management. 2014;**40**(5):1-95. DOI: 10.1177/0149206314527128

39. Paulus P. Collaborative creativity—group creativity and team innovation. In: Handbook of Organizational Creativity. Academic Press; 2012. pp. 327-357. DOI: 10.1016/B978-0-12-374714-3.00014-8. Available from: https://www.sciencedirect.com/science/article/abs/pii/B9780123747143000148

40. Romer D. Advanced Macroeconomics (Mcgraw-hill Economics) 5th ed. 2018. ISBN: 978-1260185218

41. Peri C, Lavelli V, Marjani A. Qualità nelle aziende e nelle filiere agroalimentari: gestione e certificazione dei sistemi per la qualità, per la rintracciabilità e per l'igiene. Milano: Ulrico Hoepli Editore S.p.A; 2004

42. Mariana RR, Hidayati L, Soekopitojo S. Implementing the HACCP system to the production of Bakso Malang-Indonesia. 2018;**17**(4):291-312. DOI: 10.1080/15428052.2018.1442760. Available from: https://www.tandfonline.com/doi/full/10.1080/15428052. 2018.1442760. Received 11 Oct 2017, Accepted 30 Jan 2018, Published online: 22 Mar 2018

43. Simonton DK. Creative productivity: A predictive and explanatory model of career trajectories and landmarks. Psychological Review. 1997;**104**(1): 66-89

44. Stuart T. Interorganizational alliances and the performance of firms: A study of growth and innovation rates in a high-technology industry. Strategic Management Journal. 2000;**21**(8):791-811. ISSN 1097-0266

45. Sharma S. Total Quality Management: Concepts, Strategy, and Implementation for Operational Excellence. 1st ed. Anmol Publications Pvt Ltd (January 1, 2007); 2018. ISBN-13: 978-9352805129. ISBN-10: 9352805127

46. Gopalakrishnan Nair TR, Suma V. A paradigm for metric-based inspection process for enhancing defect management. ACM SIGSOFT Software Engineering Notes. 2010;**35**(3):1. DOI:10.1145/11764810.1764827

47. Burger-Helmchen T. 2012. Entrepreneurship – Creativity and Innovative Business Models. Rijeka, Croatia: InTech. ISBN 978-953-51-0069-0. Available from: https://www.scribd.com/document/179525591/entrepreneurship-creativity-and-innovative- business-models

48. World Development Report 2005. A Better Investment Climate for Everyone. The World Bank Institute; 2005. ISBN:978-0-8213-5682-1. Available from: https://elibrary.worldbank.org/doi/abs/10.1596/0-8213-5682-8

49. Lin X-Q, Chen Y-C, Liu C-H, Li Y-Q. Measuring creativity: role of service quality management, knowledge sharing and social interaction, Total Quality Management & Business Excellence. 2023;**34**(1-2):1-18. DOI: 10.1080/14783363.2021.2021800

Author details

Enriko Ceko
Canadian Institute of Technology, Tirana, Albania

*Address all correspondence to: enrikoceko@yahoo.co.uk; enriko.ceko@cit.edu.al

References

[1] Malthus T. Essay on the Principle of Population. Oxford: Oxford University Press; 1798 (1993 printing)

[2] Ceko E. On relations between creativity and quality management culture. Creativity Studies. 2021; **14**(1):251-270. DOI: 10.3846/cs.2021.12154. Available from: https://journals.vilniustech.lt/index.php/CS/article/view/12154. file:///C:/Users/Admin/Downloads/12154-Article%20Text-53104-7-10-20210617%20(2).pdf

[3] Houghton J, Shehan P. A Primer on the Knowledge Economy. Centre for Strategic Economic Studies: Victoria University; 2000. Available from: http://docplayer.net/50330830-A-primer-on-the-knowledge-economy.html

[4] OECD. The Knowledge-Based Economy. Paris: OECD; 1996. Available from: https://www.oecd.org/officialdocuments/publicdisplaydocumentpdf/?cote=OCDE/GD%2896%29102&docLanguage=En

[5] Meusburger P, Funke J, Wunder E. In: Meusburger P, editor. Milieus of Creativity: An Interdisciplinary Approach to Spatiality of Creativity. Series: Knowledge and Space. Vol. 2. Berlin: Springer Science + Business Media B. V.; 2006. pp. 97-154

[6] Mumford MD. Where have we been, where are we going? Taking Stock in Creativity Research, Creativity Research Journal. 2003; **15**(2–3):107-120

[7] Sternberg RJ, Sternberg K, Mio J. Cognitive Psychology. Wadsworth: Cengage Learning; 2012

[8] Torrance P. Verbal Tests. Forms A and B-Figural Tests, Forms A and B. In: The Torrance Tests of Creative Thinking-Norms-Technical Manual Research Edition. Princeton, New Jersey: Personnel Press; 2008. p. 6

[9] Herrera-Usagre M. Cultural Practice, Creativity, and Innovation. In: Bericat E, Jiménez-Rodrigo M, editors. The Quality of European Societies. Social Indicators Research Series. Vol. 75. Cham: Springer; 2019. DOI: 10.1007/978-3-030-05023-8_6

[10] Leal S, Urrea J. Ingenio y Pasión. Barcelona: Lid Publishers, Inc.; 2013

[11] Sinha N. Unleash the power of creativity and innovation. International Journal of Sustainable Strategic Management. 2009

[12] Siltala R. Innovatiivisuus ja yhteistoiminnallinen oppiminen liike-elämässä ja opetuksessa. Turku: Turun Yliopisto; 2010

[13] Davila T, Epstein MJ, Shelton R. Making Innovation Work: How to Manage it, Measure it, and Profit from it. Upper Saddle River: Wharton School; 2006

[14] Ahmad N, Seymour RG. Defining Entrepreneurial Activity, OECD Report. 2006

[15] Wróbel AE, Cash P, Lomberg C. Pro-active neutrality: The key to understanding creative facilitation. Creativity and Innovation Management. 2020;**29**(3):424-437. DOI: 10.1111/caim.12372. Available from: https://onlinelibrary.wiley.com/doi/10.1111/caim.12372

[16] Drucker P. Innovation and Entrepreneurship: Practice and Principles. New York, USA: Harper Business; 1985

[17] Sarasvathy S, Dew N, Velamuri SR, Venkataraman S. Three views of entrepreneurial opportunity. In: Acs Z, Audretsch D, editors. Handbook of Entrepreneurship Research: an Interdisciplinary Survey and Introduction. New York: Springer; 2003. pp. 141-160

[18] Shane S. A General Theory of Entrepreneurship. The Individual–Opportunity Nexus. Cheltenham, UK: Edward Elgar; 2003

[19] Schumpeter JA. In: Redvers O, Elliott JE, editors. The Theory of Economic Development: An Inquiry into Profits, Capital, Credit, Interest, and the Business Cycle. New Brunswick, New Jersey. 1983. ISBN 0-87855-698-2. Available from: https://en.wikipedia.org/wiki/Innovation#cite_note-1

[20] ISO 56000:2020 (en). Innovation Management — Fundamentals and Vocabulary. International Organization of Standards; 2020

[21] Cruz-Cunha MM, Miranda IM, Gonçalves P. Handbook of Research on ICTs for Human-Centered Healthcare and Social Care Services. Vol. 2. 2013. p. 978. DOI: 10.4018/978-1-4666-3986-7. Available from: https://www.igi-global.com/book/handbook-research-icts-human-centered/72373#table-of-contents

[22] Bhasin K. This is the Difference Between 'Invention' and 'Innovation'. Business Insider; 2012

[23] Morgan J. What's the Difference Between Invention and Innovation? Forbes; 2015

[24] Schumpeter JA. Innovation is possible without anything we should identify as an invention, and the invention does not necessarily induce innovation. Business Cycles. 1939;**1**:84. Available from: https://en.wikipedia.org/wiki/Innovation

[25] Hidajat TJ, Hendarman AF. Relationship among soft skills, hard skills, and innovativeness of knowledge workers in the knowledge economy era. In: 10th Triple Helix Conference 2012. Procedia - Social and Behavioral Sciences. Vol. 52. 2012. pp. 35-44 1877-0428. Available from: https://moam.info/relationship-among-soft-skills-hard-skills-and-59cecf501723dd7cffa40067.html

[26] Utterback J. The process of technological innovation within the firm. Academy of Management Journal. 1971; **14**(1):78. Available from: https://en.wikipedia.org/wiki/Innovation#cite_note-1

[27] Drucker P. The Discipline of Innovation. Harvard Business Review. 2002. Available from: https://en.wikipedia.org/wiki/Innovation#cite_note-1

[28] Shalley CE, Hitt MA, Zhou J. The Oxford Handbook of Creativity, Innovation, and Entrepreneurship. Oxford University Press; 2015. ISBN 978–0–19–992767–8

[29] Perry-Smith J, Mannucci PV. Social networks, creativity, and entrepreneurship. In: The Oxford Handbook of Creativity, Innovation, and Entrepreneurship. Oxford University Press; 2015. ISBN 978–0–19–992767–8

[30] Byron K, Khazanchi S. Rewards' relationship to creativity, innovation, and entrepreneurship at the Oxford handbook of creativity, innovation, and entrepreneurship. In: The Oxford Handbook of Creativity, Innovation, and Entrepreneurship. Oxford University Press; 2015. ISBN 978–0–19–992767–8

[31] Gilson LL, Lim HS, Litchfield RC, Gilson PW. Entrepreneurial creativity:

The role of learning creativity in teams: A key building block for innovation and entrepreneurship. In: The Oxford Handbook of Creativity, Innovation, and Entrepreneurship. Oxford University Press; 2015. ISBN 978–0–19–992767–8

[32] Shin SJ. Leadership and creativity: The mechanism perspective. In: The Oxford Handbook of Creativity, Innovation, and Entrepreneurship. Oxford University Press; 2015. ISBN 978–0–19–992767–8

[33] Leydesdorff L. The Triple Helix of University-Industry-Government Relations. The Netherlands: The University of Amsterdam; 2012. Available from: http://www.leydesdorff.net/th12/th12.pdf

[34] Tjakraatmadja JH, Wicaksono A, Martini L. Institut Teknologi Bandung (ITB) is a Potential knowledge hub to create value from academia, business, and government linkages. In: Beyond The Knowledge Trap. World Scientific Publishing; 2011

[35] Global Innovation Report. Preface vii. 13th ed. World Intellectual Property Organization; 2020. Available from: https://www.wipo.int/edocs/pubdocs/en/wipo_pub_gii_2020.pdf

[36] Harrington HJ, Mathers DD. ISO 9000 and Beyond: From Compliance to Performance Improvement. New York: McGraw-Hill; 1997

[37] Ceko E. Total Quality Management. Wisdom University Press; 2017

[38] Robbins SP. Organizational Behavior was 12th. New York: Prentice Hall; 1999

[39] Schein EH. 2020 Organizational Culture and Leadership. San Francisco: Jossey-Bass Publishers; 2013

[40] Harvey L, Green D. Defining quality. Assessment, and Evaluation in Higher Education. 1993;**18**(1):9-34

[41] Vlăsceanu L, Grünberg L, Pârlea D. In: Seto M, Wells PJ, editors. Quality Assurance and Accreditation: A Glossary of Basic Terms and Definitions. Bucharest: UNESCO-CEPES; 2007

[42] EUA - European University Association. Quality Culture in European Universities: A Bottom-Up Approach. Report on the Three Rounds of the Quality Culture Project; 2002 – 2006. ISBN: 9-0810-6983-7

[43] Gordon G, Owen C. SHEEC on Management of Quality: Cultures of Enhancement and Quality Management Systems and Structures. 2008. Available from: https://www.enhancementmusic.UK/docs/report/-management-of-quality-cultures-of-quality-enhancement.pdf?sfvrsn=12

[44] Harvey L, Stensaker B. Quality culture: Understandings boundaries, and linkages. European Journal of Education: Research, Development, and Policy. 2008;**43**(4):427-442

[45] Fernald LD. Perspectives Archived 8 June 2020 at the Wayback Machine. Thousand Oaks, CA: Sage Publications; 2008. pp. 12-15

[46] Hockenbury & Hockenbury. Psychology. Worth Publishers; 2010

[47] Bureau of Labor Statistics, U.S. Department of Labor. Occupational Outlook Handbook. 11th ed. 2010

[48] Truxillo DM, Bauer TN, Erdogan B. Psychology and Work: Perspectives on Industrial and Organizational Psychology. New York: Psychology Press; 2016. ISBN 1134705697

[49] Koppes LL. Industrial-Organizational Psychology. In: Weiner, editor. Handbook of Psychology. Vol. 1. History of Psychology; 2003

[50] Myers. Motivation and Work. Psychology. New York, NY: Worth Publishers; 2004

[51] Clinton AB. The Importance of Innovation to Global Psychology. 2020. Available from: https://www.apa.org/international/global-insights/innovation [Retrieved 15th Jan 2023]

[52] Jones K. The Psychology of Innovation: From Purpose to Gratitude. Forbes Agency Council; 2016

[53] Kumar M, Bharadwaj A. Psychology of innovation: Innovating human psychology? In: Gatzweiler FW, von Braum J, editors. Technological and Institutional Innovations for Marginalized Smallholders in Agricultural Development. Springer; 2016. pp. 65-80

[54] Drevitch G. The Surprising Psychology of Innovation. 2021. Available from: https://www.psychologytoday.com/intl/blog/long-fuse-big-bang/202107/the-surprising-psychology-innovation

[55] Amabile TM. The social psychology of creativity: A componential conceptualization. Journal of Personality and Social Psychology. 1983b;**45**:357-376

[56] Randel AE, Jaussi KS. Leading for uniqueness: The role of uniqueness in facilitating creativity in employees' self-concepts. In: Karwowski M, Kaufman JC, editors. The Creative Self: Effect of Beliefs, Self-Efficacy, Mindset, and Identity. Elsevier Academic Press; 2017. pp. 289-299. DOI: 10.1016/B978-0-12-809790-8.00016-9

[57] An D, Song Y, Carr M. A comparison of two models of creativity: Divergent thinking and creative expert performance. Personality and Individual Differences. 2016;**90**:78-84. DOI: 10.1016/j.paid.2015.10.040

[58] Runco MA, Jaeger GJ. The standard definition of creativity. Creativity Research Journal. 2012;**24**(1):92-96. DOI: 10.1080/10400419.2012.650092

[59] Csikszentmihalyi, Sawyer. Shifting the focus from individual to organizational creativity. In: The Systems Model of Creativity. 2014. pp. 67-71. DOI: 10.1007/978-94-017-9085-7_6

[60] Oldham GR, Cummings A. Employee creativity: Personal and contextual factors at work. Academy of Management Journal. 1996;**39**(3):607-634. DOI: 10.2307/256657

[61] Amabile TM, Conti R, Coon H, Lazenby J, Herron M. Assessing the work environment for creativity. Academy of Management Journal. 1996;**39**:1154-1184. DOI: 10.5465/256995

[62] Judd DK. The psychology of quality. Total Quality Management. 1995;**6**(3):287-291. DOI: 10.1080/09544129550035459 [Published online: 25 Aug 2010]

[63] Jones T, Buch KK. Total quality management. In: The SAGE Encyclopedia of Industrial and Organizational Psychology. 2nd ed. 2017. DOI: 10.4135/9781483386874

[64] Ceko E. On relations between innovation, creativity and quality management. In: 10th Mediterranean International Conference. Tenerife, Spain: University San Cristobal de La Laguna; 2022

[65] Ceko E. A new approach to evaluate the index of ISO 9001. Socioeconomic

Challenges. 2022;**6**(3):5-22. DOI: 10.21272/sec.6(3).5-22.2022

[66] Henderson RM, Clark KB. Architectural innovation: The reconfiguration of existing product technologies and the failure of established firms. Administrative Science Quarterly. 1990;**35**(1):9. DOI: 10.2307/2393549

[67] Nickerson RS. Developing Intelligence Through Instruction

[68] Wróbel AE. Creative Facilitation in Innovation Management: Facilitator's Role in Shaping Team Processes. The Technical University of Denmark; 2019

[69] Creativity Index. Marten Prosperity Institute, University of Toronto; 2020

[70] The relationship between undernutrition and behavioral development in children. Journal of Nutrient. 1995;**125**:2211S-2284S. Copyright (c) 1995 by the American Institute of Nutrition

Section 3

Other Worlds for Creativity and Innovation

Chapter 8

The Life^{2}Well Project: Investigating the Relationship between Physiological Stress and Environmental Factors through Data Science, the Internet of Things and Do-it-Yourself Wearables

Nguyen Duc Minh Anh, Nguyen Thien Minh Tuan, Kenneth Y.T. Lim and Ahmed H. Hilmy

Abstract

This chapter reports a study conducted by students as an independent research project under the mentorship of a Research Scientist at the National Institute of Education, Singapore. In the Life^{2}Well Project (Learning at the intersection of AI, physiology, EEG, our environment and well-being) identical units of a wearable device containing environmental sensors (such as ambient temperature, air pressure, infrared radiation and relative humidity) were designed and worn respectively by five adolescents from July to December 2021. Over the same period, data from these sensors was complemented by that obtained from smartwatches (namely blood oxygen saturation, heart rate and its variability, body temperature, respiration rate and sleep score). More than 40,000 data points were eventually collected, and were processed through a random forest regression model, which is a supervised learning algorithm that uses ensemble learning methods for regression. Results showed that the most influential microclimatic factors on biometric indicators were noise, and the concentrations of carbon dioxide and dust. Subsequently, more complex inferences were made from Shapley value interpretation of the regression models. Such findings suggest implications for the design of living conditions with respect to the interaction of microclimate and human health and comfort.

Keywords: physiological response to microclimate, random forest regression analysis, the internet of things, citizen science, environmental sensor, wearables

1. Introduction

Climate change has been one of the most urgent problems to confront in the 21st century. The fifth report of the Intergovernmental Panel on Climate Change

(IPCC) confirms that human influence on the climate system is clear and growing, with impacts observed across all continents and oceans [1]. Climate change has been degrading the quality of life for every creature on Earth. Glaciers have shrunk, ice on rivers and lakes is breaking up earlier, plant and animal ranges have shifted and trees are flowering sooner [2].

From the United Nations World Urbanization Prospects in 2017, 4.1 billion people were living in urban areas [3]. This means over half of the world (56% in 2020) live in urban settings. Urbanization is happening rapidly and it also accounts for global climate change. The rapid and large-scale urbanization leads to severe land-use conversion and impacts ecosystem services [4]. The latter refer to the direct and indirect contributions of ecosystems to human well-being. With ecosystems being damaged, human well-being is also affected. In the context of this pressing problem, urban microclimate studies have been gaining prominence due to rapid urbanization [5].

A microclimate is a small area within the surrounding larger area with a different climate [6]. Any given climatic region therefore comprises many other types of microclimates, which vary in characteristics from the region as a whole. Because our planet in general is broadly conducive to life, we – as humans – have populated its land masses. Comparing the human scale to that of the various habitats in which we live, the difference of these scales means that changes in the climates of these habitats may disproportionately affect the conduct of our daily activities.

That is the reason why the authors felt compelled to explore the relationships between microclimate and our physical and mental well-being. It is self-evident that climate change has various effects on the well-being of a person. As human beings we are conscious and aware of our surroundings, and our responses to changes in microclimate may affect our emotions. To elaborate, climate change might precipitate changes to micro-climates to the extent that for those inhabiting these biomes, the changes might be detrimental to physical and mental well-being. For instance, a study by Liu et al. [7] concluded that "the increasing research interest in thermal comfort and health has heightened the need to figure out how the human body responds, both psychologically and physiologically, to different microclimates".

This chapter reports a study conducted by students as an independent research project under the mentorship of a Research Scientist at the National Institute of Education, Singapore. In the Life^{2}Well Project (Learning at the intersection of AI, physiology, EEG, our environment and well-being) identical units of a wearable device containing environmental sensors (such as ambient temperature, air pressure, infrared radiation and relative humidity) were designed and worn respectively by five adolescents from July to December 2021. Over the same period, data from these sensors was complemented by that obtained from smart-watches (namely blood oxygen saturation, heart rate and its variability, body temperature, respiration rate and sleep score).

Among others, our work was inspired by earlier work of [8], in which they concluded that modified urban microclimates have a deep impact on the comfort of inhabitants. Palme and Salvati lamented that while there have been various studies conducted on the effects of urbanization on microclimate and how microclimate has changed human health in efforts to redesign and restructure urban areas., there have been relatively fewer studies on the relationships between microclimate and human health and emotions.

We see one of the potential contributions of our work to be our use of self-designed, low-cost, wearable units for measuring microclimate. The relatively low cost of these wearables has positive implications on the affordance of scalability, and – consequently – on crowd-sourced citizen science in this as yet under-reported field of the relationships between microclimate and well-being.

2. Review of literature

2.1 The importance of microclimate

Microclimate has been defined as the suite of climatic conditions measured in localized areas near the earth's surface [9]. Microclimate includes environmental variables such as temperature, light, wind speed, and moisture. It has been critical throughout human history, providing meaningful indicators for habitat selection and other activities [10]. Regardless of the global biomes in which we live, it is specifically microclimate that our bodies respond to, and not to the descriptors of the respective climatic region as a whole. For example, farmers have long used localized seasonal changes in temperature and precipitation to schedule their agricultural activities. Microclimate directly influences ecological processes and reflects subtle changes in ecosystem function and landscape structure across geographical scales [11].

As an example related to human health, microclimate in urban areas affects our thermal comfort [8]. However, relationships between microclimate and biological processes are complex and often nonlinear. For example, plant distribution can be perceived as a function of light, temperature, moisture, and vapor deficit [10]. Therefore, just a subtle change in microclimate could cause detrimental effects on human emotions and health, aside from just thermal comfort.

2.2 Effects of urbanization on microclimate

Rapid urbanization, especially in developing countries, has led to large flows of migration to urban areas [12]. According to a 2021 report authored by the market research firm Statista, the degree of urbanization worldwide was at around 56% in 2020 [13]. With rapid urbanization, changes to the urban environment and climate are inevitable [14]. The rate of urbanization is very high and anthropogenic effects on the climate of Earth are difficult to predict.

At local scales, activities associated with land use and land cover changes and urbanization induce impacts such as changes in atmospheric composition in water and energy balances and changes in the ecosystem [15]. By definition, ecosystems are interconnected, therefore a small change in any component can result in non-linear effects elsewhere. For example, according to a study conducted by Xiong et al. [16] in 2015 on the influence of different air temperature step-changes on human health and thermal comfort, perspiration, eye-strain, dizziness, accelerated respiration and heart rate were all sensitive self-reported symptoms.

Due to global climate change and intensifying urban heat island effects, urban living environments have deteriorated, becoming increasingly detrimental to human thermal comfort and health, not only psychologically, such as in terms of thermal sensation, mood, and concentration, but also physiologically by way of, for example, sunburn, heat stroke, and heat cramps [7]. They also cautioned that "global climate change and intensifying heat islands have reduced human thermal comfort and health in urban outdoor environments". Other studies on re-naturing cities have found that changes to urban microclimate can potentially exacerbate the risks of meteorological hazards such as heatwaves. Heat-related issues pose an impact not only to the environment but also lead to heat-related human health problems and in extreme cases, cause deaths [15]. This is only one aspect of how microclimate can affect humans.

3. Materials and methods

3.1 Collecting microclimate data

To collect microclimate data, a small portable device that could be worn on the waist was designed in order to measure the following ambient environmental conditions:

- Noise level
- Infrared radiation through light intensity
- The amount of dust
- Carbon dioxide concentration
- Temperature
- Relative humidity
- Air pressure.

In total, five such devices were built by the authorial team. Each device was worn by the first and second authors and by their adolescent peers. The devices measured 12 cm by 6 cm by 2 cm respectively, and contained a low-cost sensor for each of the preceding environmental variables listed, together with a battery with sufficient capacity to power the sensors over the course of a typical day. The device could be secured to a belt by two regular clothes clips.

Every 15 minutes, each device would automatically log its measurements in to a designated cloud-based spreadsheet. At the same time, the device would ping the nearest publicly-accessible weather station (as provided by the Singapore Meteorological Service) for the wind speed and wind direction prevailing at that time. **Figure 1** shows schematics of the components, and **Figure 2** shows an assembled device.

3.2 Collecting data on mental state and health

Biometric data was collected using three Huawei Honor Band 6 smartwatches and two Fitbit Sense smartwatches. Each smartwatch measured:

- SpO2: the oxygen saturation in blood, which is the amount of oxygen-carrying hemoglobin in the blood relative to its capacity; and
- Heart rate.

These two variables were used by both types of watches to generate a so-called stress score (the latter pertinent to waking moments and arbitrarily indexed from 1 to 100).

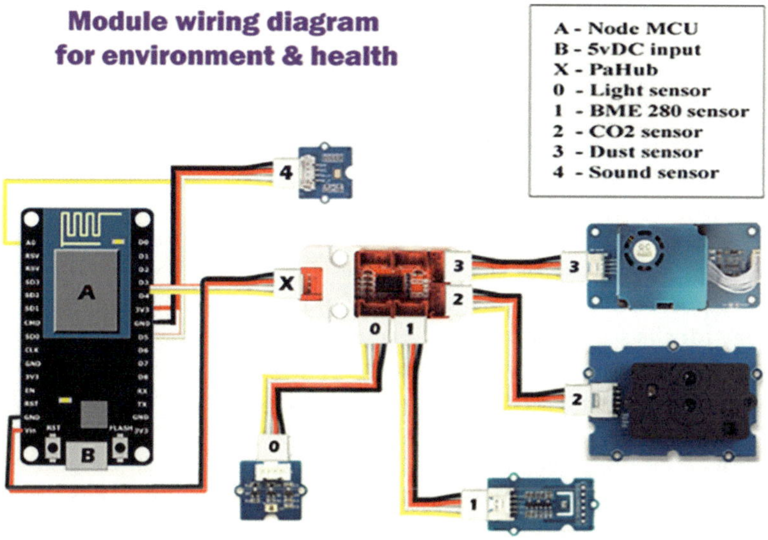

Figure 1.
Schematic of sensors for the wearable.

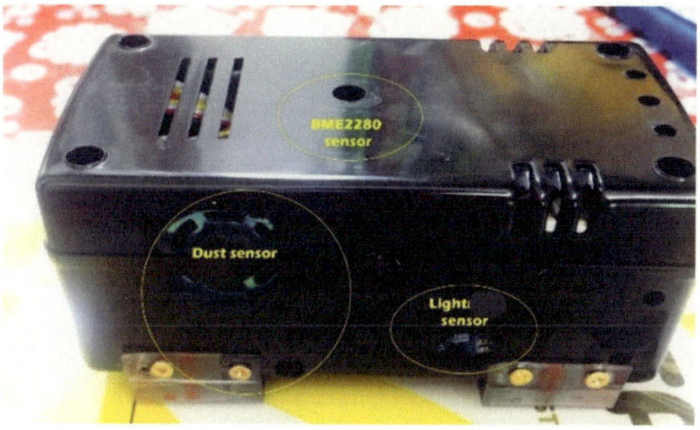

Figure 2.
Assembled wearable device..

In addition, the Fitbit Sense watches were able to measure:

- Skin temperature at the wrist (indicated as degrees Celsius off the baseline of core body temperature);
- Respiration rate (measured in breaths per minute); and
- Heart rate variability, which is the physiological phenomenon of variation in the time interval between heartbeats.

Heart rate variability was – in turn – indicated through the variables of:

- Root mean square of successive differences between successive beat-to-beat intervals (rMSSD)

- Low frequency, which indicates long-term variations in heart rate and reflects activity from both the sympathetic and parasympathetic branches of the autonomic nervous system; and

- High frequency, which indicates short-term variations in heart rate and reflects activity from the parasympathetic branch.

Finally, the Fitbits used the duration of deep sleep, resting heart rate, and restlessness while supine, to suggest a sleep quality score (indexed from 1 to 100). According to Fitbit, this sleep score takes into account sleep duration, sleep quality, and restoration.

As a proxy for the state of mind of the five participants during their waking hours, they voluntarily self-reported against a ten-point Likert scale twice a day, once in the morning and the other before sleep. In earlier work by [17], it has been recognized that there is a high degree of connectivity among the neural structures of the brain and its systems. Emotions and cognition, although having separate features and influences, are dialectic, integrated and co-mingled in the brain. It has been suggested that emotions play a central role in the evolution of consciousness, influence the emergence of higher levels of awareness and largely determine the content and focus of consciousness throughout one's lifespan [17]. As such, by attempting to explore the connections between participants' feelings and corresponding biometric data, we can perhaps glimpse how participants respond physiologically and affectively when they are in a specific microclimate.

Given this background, the self-report online form consisted of four questions. Two questions invited the five participants to report their emotional and physical health based on prevailing microclimate at the time, against a ten-point Likert scale, while the other two questions invited them to elaborate on their respective responses in prose.

4. Findings

In total 43,579 data entries are collected from July to December 2021; after cleaning, 43,434 data points were used. Reference values for carbon dioxide concentration are approximately 400 ppm, for PM2.5 dust 10 μg/m3 annual mean, and atmospheric pressure at sea-level is 1013.25 millibars (see, for example, [18–20]. From historical records, the diurnal temperature range of Singapore is 25 deg. C to 33 deg. C; relative humidity in the island nation ranges from 60–90% typically. From June to September, the climate of Singapore is influenced by the southwest monsoon, after which is the inter-monsoonal period of relatively weaker winds.

Data was analyzed using the Python libraries Sklearn and SHAP. Correlations were drawn based on a best-fit line graph between each respective microclimatic and biometric/well-being variable.

Random forest regression models were then performed. Random forest regression are supervised learning algorithms that use ensemble learning methods for regression. In turn, an ensemble learning method is a technique that combines predictions from multiple machine learning algorithms to make a more accurate prediction than a single model; in the case of the study reported, it used multiple regression models. Finally, outlier events were identified for subsequent investigation if necessary.

To measure the respective contributions of the various predictors (the microclimate variables) against the actual values, Shapley summary plots were generated from the training data. Shapley values can be thought of as the average of the marginal contributions across all possible permutations within a given model. Simply put, Shapley values decompose a prediction to show the impact of each feature, by showing how much each feature contributed to the overall predictions.

The sections below elaborate on the respective contributions of each of the microclimatic variables measured in this study to selected biometric indicators and indicators of well-being of interest. Within each section, a Shapley summary plot showing the contributions of the various microclimate variables to the given biometric/well-being indicator is presented.

In the Shapley plots, variables are ranked in descending order of importance, and the situation of each dot along the x-axis shows whether the effect of that value is associated with a higher or lower prediction. Simple exponential smoothing was used on the models. Each environmental feature had its smoothed counterpart with the format: "(feature)_", which was derived from values of that feature from preceding hours, with older values being exponentially less important than the current value. For example: for f(t), we have f(t-1), f(t-2), ... affecting it, but f(t-1) will be of more importance to f(t) than f(t-2), f(t-3),... exponentially.

The features "0", "1", ... "23" represent the hours of the day that data points were recorded, while "31" to "37" are days of the week (Monday to Sunday), which correspond to the time of recording of those data points. These features are binary, so 1 or high means true and 0 or low means false. For example, if a data point was recorded on 6 AM on a Monday then "6" and "31" would be 1 and the remaining binary features would be 0. Finally, the color shows whether that variable is high (red) or low (blue) for that observation. In this way, Shapley summary plots combine feature importance with feature effects.

5. Associative relationships of microclimate and physical health

5.1 Heart rate

Figure 3 shows the Shapley Summary Plot of the associations of microclimate with the variable 'heart rate'.

From **Figure 3**, the variable 'heart rate' is most strongly associated with the variable 'carbon dioxide concentration' (feature importance of 0.15), followed by the variable 'sound' (feature importance of 0.14) and by three variables which affects heart rate co-equally (feature importance of 0.08), namely 'ambient temperature', 'dust concentration' and 'air pressure'. From the above Shapley Summary Plot, the lower the carbon dioxide concentration, the lower the heart rate. As for sound, the higher the sound in the surrounding microclimate is, the higher the heart rate level. Similarly, the higher the pressure and dust concentration are, the higher the heart rate level while heart rate value decreases when ambient temperature increases. An interesting observation is that heart rate dropped from 5 pm onward.

5.2 Heart rate variability (rMSSD)

In the context of heart rate variability, rMSSD refers to the physiological phenomenon of variation in the time interval between heartbeats. **Figure 4** shows the Shapley Summary Plot of the associations of microclimate with this variable.

From **Figure 4**, the variable 'root mean square of successive beat-to-beat interval differences' ('rMSSD') is most strongly associated with the variable 'sound' (feature

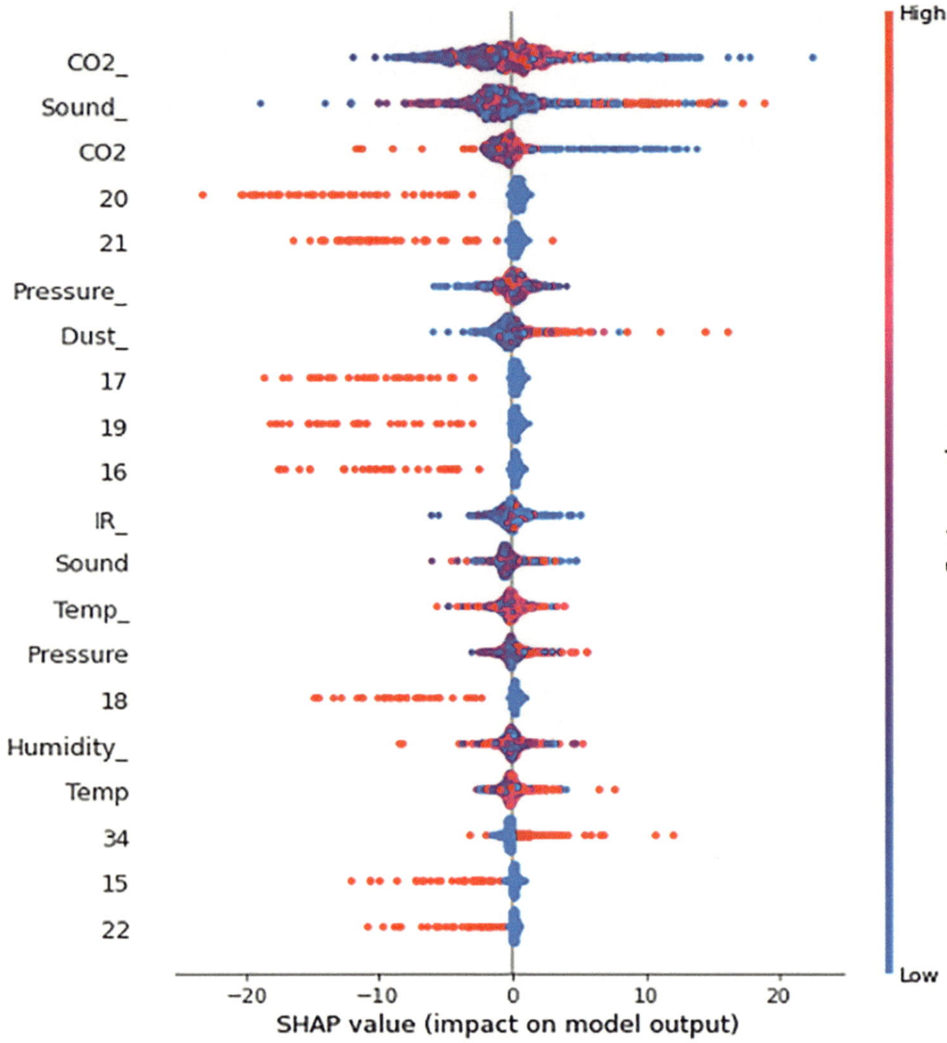

Figure 3.
Shapley summary plot of microclimatic factors and heart rate.

im-portance of 0.46), followed by the variable 'dust concentration' (feature importance of 0.09), then co-equally (feature importance of 0.07) by the variables 'ambient temperature' and 'wind direction'. From the above Shapley Summary Plot, the higher sound is, the lower the rMSSD is. The higher dust concentration is, the lower the rMSSD. Similarly, the higher the ambient temperature and the more westerly the wind, the lower the rMSSD.

5.3 Skin temperature at the wrist

Figure 5 shows the Shapley Summary Plot of the associations of microclimate with the variable 'skin temperature at the wrist'.

From **Figure 5**, the variable 'skin temperature at the wrist' is most strongly associated with the variable 'sound' (feature importance of 0.21), followed by the variable 'dust concentration' (feature importance of 0.13) and the variable 'ambient

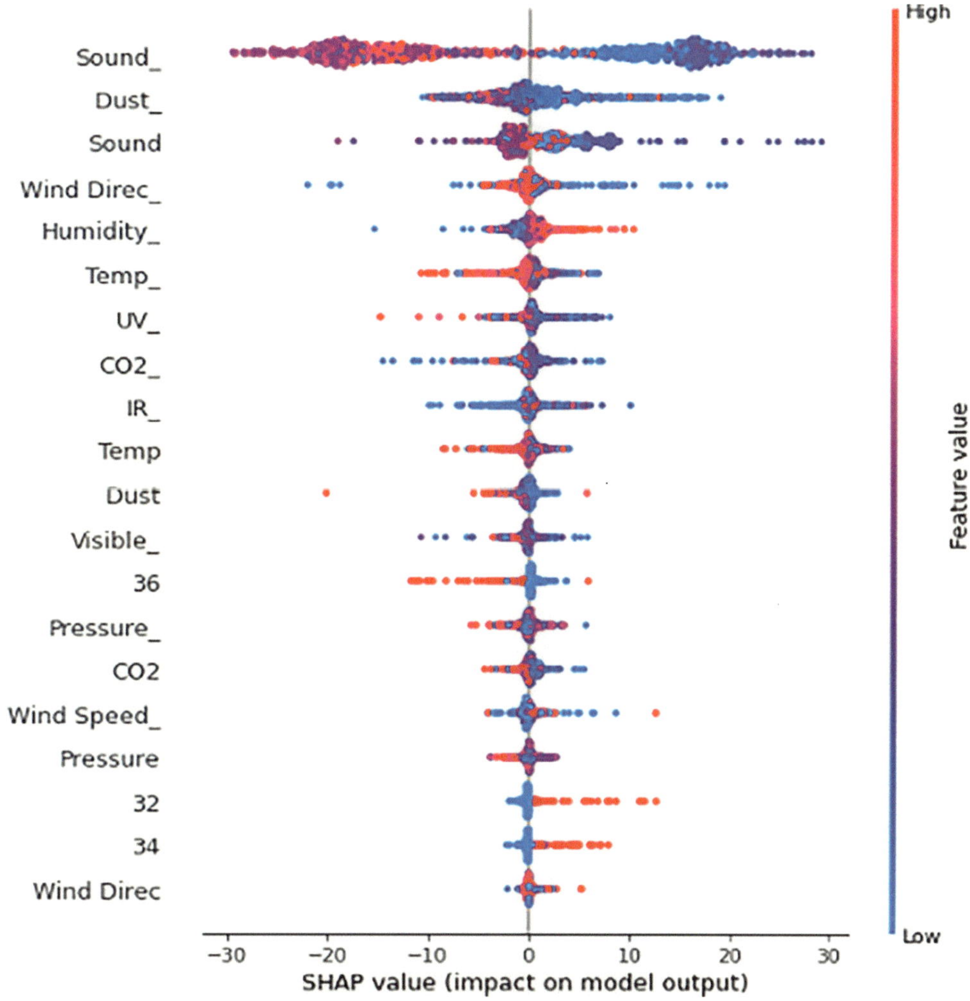

Figure 4.
Shapley summary plot of microclimatic factors and rMSSD.

temperature' (feature importance of 0.12). From the above Shapley Summary Plot, the higher sound is, the higher the skin temperature. In attempting to interpret this, one might consider sound to be a proxy indicator of the nature of one's immediate environment, which – in turn – may have attributes which results in physiological responses, including with respect to skin temperature. The lower amount of dust concentration present is, the higher the skin temperature. The higher ambient temperature is, the higher the skin temperature but it rises at a lower rate.

5.4 Self-reported physical state

Figure 6 shows the Shapley Summary Plot of the associations of microclimate with the variable 'self-reported physical state'.

From **Figure 6**, the variable 'self-reported physical state' is most strongly associated with the variable 'sound' (feature importance of 0.17), followed by the variable 'carbon dioxide concentration' (feature importance of 0.15) and by the variable 'dust

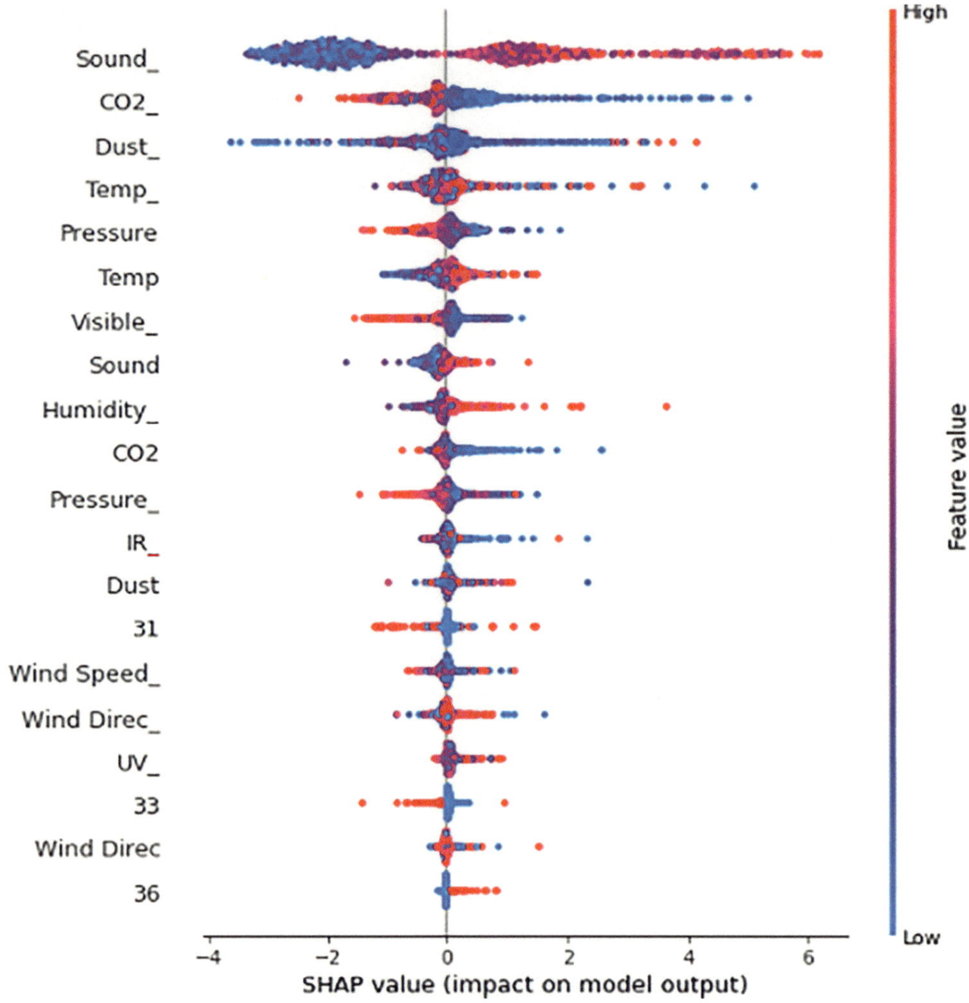

Figure 5.
Shapley summary plot of microclimatic factors and skin temperature.

concentration' (feature importance of 0.13). From the above Shapley Summary Plot, the higher level of sound, the higher the value of self-reported physical state. Similarly, the higher carbon dioxide concentration is, the higher the value of self-reported physical state. Finally, as dust concentration decreases, the value of self-reported physical state increases.

6. Associative relationships of microclimate and mental health

6.1 Stress score

The variable 'stress score' is generated by the smartwatches and is derived from oxygen saturation and heart rate. **Figure 7** shows the Shapley Summary Plot of the associations of microclimate with this variable.

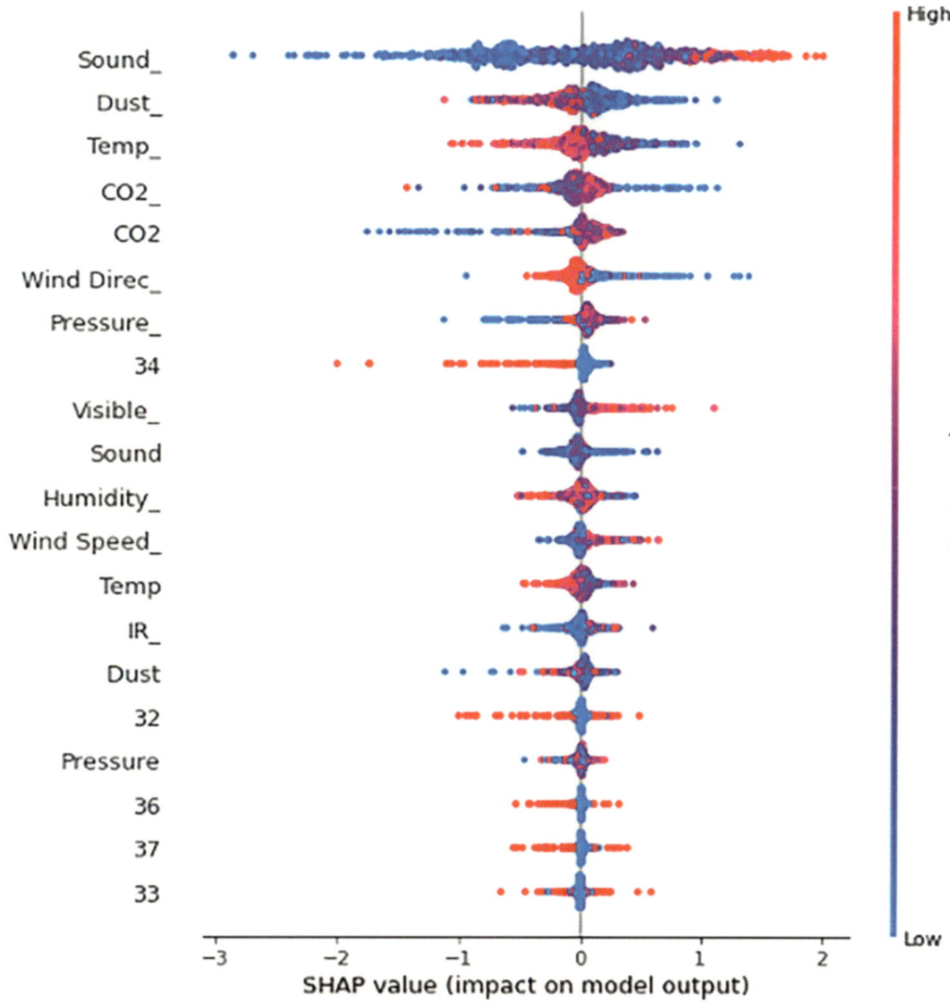

Figure 6.
Shapley summary plot of microclimatic factors and self-reported physical state.

From **Figure 7**, the variable 'stress score' is most strongly associated with the variable 'infrared radiation' (feature importance of 0.363), followed by the variable 'carbon dioxide concentration' (feature importance of 0.226) and by the variable 'ambient temperature' (feature importance of 0.104). The higher the infrared radiation, the higher the stress level. As for carbon dioxide concentration, the higher the concentration, the higher the stress level is. For ambient temperature, the lower the temperature, the lower the stress level is.

6.2 Self-reported mental state

Figure 8 shows the Shapley Summary Plot of the associations of microclimate with the variable 'self-reported mental state'.

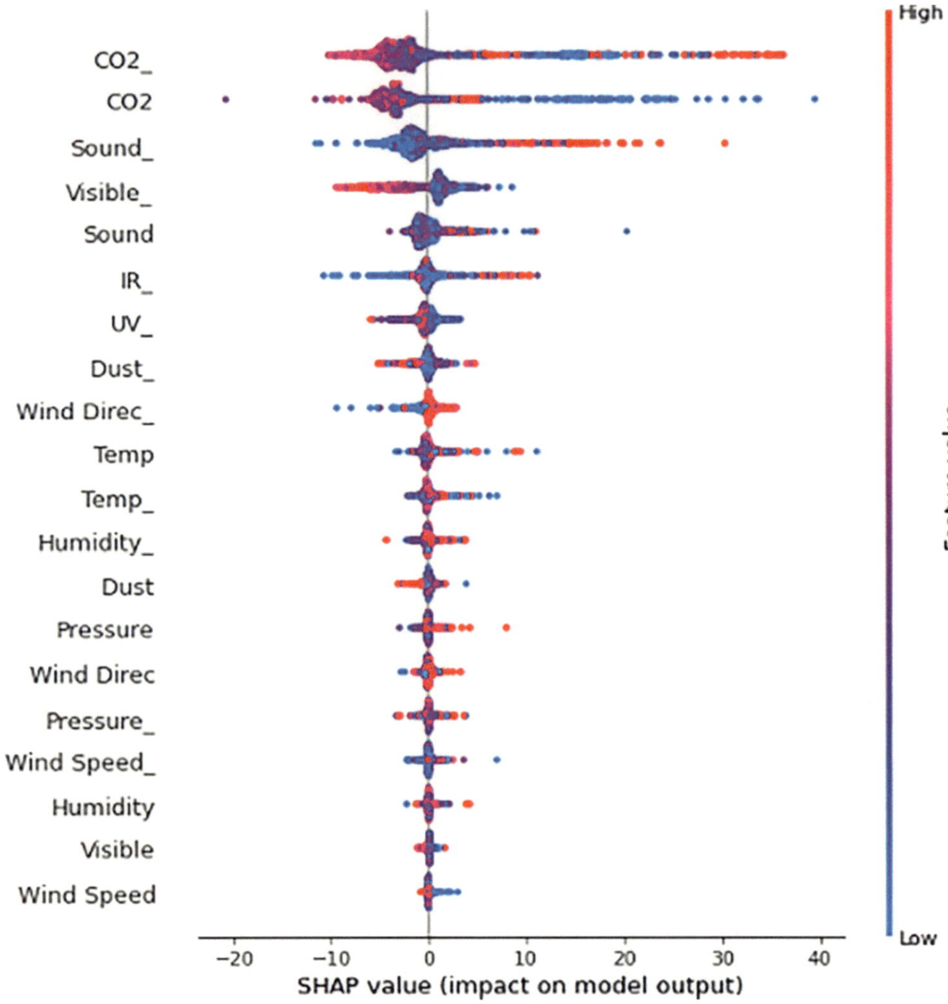

Figure 7.
Shapley summary plot of microclimatic factors and stress score.

From **Figure 8**, the variable 'self-reported mental state' is most strongly associated with the variable 'sound' (feature importance of 0.18), followed by the variable 'carbon dioxide concentration' (feature importance of 0.17) and by the variable 'dust concentration' (feature importance of 0.13). From the above Shapley Summary Plot, the higher level of sound is, the higher the value of self-reported mental state. Similarly, as carbon dioxide concentration increases, self-reported mental state increases. Finally, self-reported mental state value decreases as dust concentration increases.

7. Discussion

The preceding analysis suggests that the more significant microclimatic factors are sound, carbon dioxide concentration and dust concentration. These factors were continually associated with high feature importance scores in both physical and

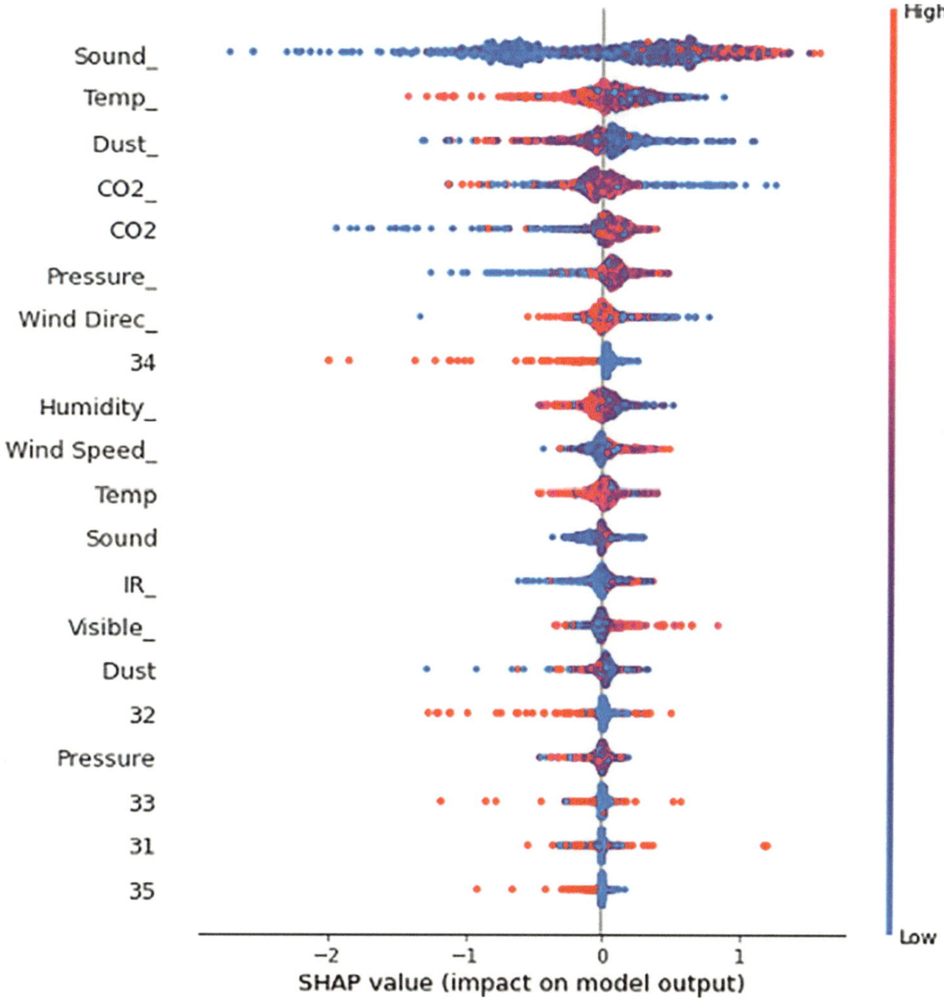

Figure 8.
Shapley summary plot of microclimatic factors and self-reported mental state.

mental well-being measures, and in both objective scores recorded from electronic instruments and more subjective self-report forms.

We acknowledge that the true influence of each value is much more abstract than the assigned Shapley value due to the 'black box' nature of random forest regression; further we also acknowledge that model interpretability does not mean causality. Notwithstanding these caveats, we have chosen to share the Shapley plots in this paper because they allow us to examine the contributions of features in ways that can be easily understood and used to make inferences.

The associations reported in this paper bear out causal linkages reported by similar studies in the field. For example, Kessel et al. [21] have documented positive relationships between ambient temperature and both body core temperature as well as corneal temperature, and Che Muhamed et al. [22] have documented positive relationships between relative humidity and body core temperature.

As for heart rate, Randall and Shelton [23] found that "carbon dioxide excess causes an increase in ventilation volume by virtue of a greater depth of breathing, the frequency decreasing slightly. The heart rate goes up with increasing carbon dioxide concentrations". Verberkmoes et al. [24] have shared that "the influence of atmospheric pressure and temperature on the incidence of acute aortic dissections may be explained by an increase in sympathetic activity, which is responsible for higher blood pressure, and heart rate". Another factor that affects heart rate is dust concentration. As suggested in research by Pope et al. [25], elevated particulate levels were associated with increased mean heart rate, and decreased overall heart rate variability. Sound also plays a part in affecting heartbeat, according to past research – for example, [26] – the higher the noise level, the higher the heart rate. Heart rate drops at night when humans are sleeping. According to Ahmed [27], "a balance of impulse from the sympathetic and the parasympathetic nerves determine a person's baseline heart rate. Interestingly, in experiments where a person's nerve supply is blocked, the heart rate is often higher; this would suggest that the parasympathetic nerve impulses that serve to slow the heart rate down are the predominant force under normal resting conditions. This is particularly evident at night when most people have a significant drop in heart rate".

Heart rate variability has been shown to be influenced by carbon dioxide concentration [28]. Our findings are congruent with studies which have elaborated on how and why heart rate variability is decreased with higher dust concentration [29], and with higher noise levels [30].

In terms of the associative relationships between microclimate and mental health, Mullins and White [31] have observed that "we find that higher temperatures increase emergency department visits for mental illness, suicides, and self-reported days of poor mental health". In a similar paper by Sygna et al. [32], results suggested that "individuals with poor sleep quality may be more vulnerable to effects of road traffic noise on mental health than individuals with better sleep quality". This position is congruent with that of the present study, in which extremes of sound levels were associated with both lower mental well-being and lower sleep quality (the latter as measured through sleep score).

As the present paper is inspired as a response to anthropogenic climate change, the extent to which carbon dioxide concentrations affect health and well-being is of interest. Kajtár et al. [33] have reported that well-being - as well as capacity to focus attention - both decline when carbon dioxide concentration in the air increases nearly tenfold to 3000 ppm.

The preceding parallels between earlier studies and the present study are encouraging, because a foundational paradigm driving our work was that of citizen science and the democratization of small-scale, low-cost research as enabled by data science and the Internet of Things (IoT). We assembled the wearables from off-the-shelf parts and coded them ourselves. Technology has enabled such democratization not only in terms of the (relatively) low cost of hardware and the open source movement in general, but also in terms of recent developments in data science and machine learning. The latter have meant that the large and burgeoning datasets associated with the use of IoT can be accessible and intelligible to wider cohorts of students and researchers. The parallels be-tween our work and earlier studies suggest that – going forward, in a world where anthropogenic climate change is a (sad) reality – meaningful scientific and geographic inquiry can be undertaken by a much wider cross-section of the general public than was once previously possible.

Several opportunities therefore suggest themselves for potential future work, which could take the form of either scaling up or translation to investigate other microclimatic variables (such as the role of infrared radiation on well-being), biometric indicators (such as readings from electroencephalograms) and socio-demographic contexts. These and other possible avenues for future work will go some way to addressing the limitations of the present study, foremost among which was the movement restrictions associated with the necessity to follow COVID-19 protocols throughout the duration of the study. These movement restrictions meant that the microclimates sampled were necessarily limited in variety. It is hoped that possible future extensions of this study in 2022 may not be as constrained.

8. Concluding remarks

The study reported in this paper set out to investigate the associative relationships between microclimate and physiological responses and well-being. We approached the investigation from the perspective of citizen science, conceptualizing, designing and fabricating what we could. We analyzed the resulting datasets informed by contemporary understandings of data science and machine learning.

Our results suggest that sound level, carbon dioxide concentration and dust con-centration feature more importantly in regression models trained on our datasets. These findings are congruent with preceding studies, and we see a primary contribution of our work as the demonstration that – in an age of anthropogenic climate change – broader cohorts of students, researchers, and the general public have potential access to tools, methods and means of analyses that were once deemed only within the reach of a privileged few due to reasons of cost, fragility, and complexity.

It is our hope that our study contributes in a small way to a body of work to help urban planners, designers of living spaces, caregivers – among many others – conceptualize modes of human habitation in sympathy with the needs of our planet, in an era when humankind has more potential than it has ever had to influence biomes in general for better or for worse.

In conclusion, it is important to seek to understand the complex and non-linear relationships between microclimate, physiological response and well-being. This paper has reported a study conducted by students in which wearable devices containing environmental sensors were designed and worn from July to December 2021. Over the same period, data from these sensors was complemented by that obtained from smartwatches. We hope to contribute to the body of literature on the relationships between microclimate and well-being. It is our hope that this wider body will catalyze subsequent studies to address health problems arising from climate change, in order to boost productivity and life satisfaction in the not too distant future.

Author details

Nguyen Duc Minh Anh[1], Nguyen Thien Minh Tuan[1], Kenneth Y.T. Lim[2*] and Ahmed H. Hilmy[2]

1 Independent Author, Singapore

2 National Institute of Education, Nanyang Technological University, Singapore

*Address all correspondence to: kenneth.lim@nie.edu.sg

References

[1] Pachauri RK, Meyer LA. Climate Change 2014: Synthesis Report. Contribution of Working Groups I, II, III to the Fifth Assessment Report of the Intergovernmental Panel on Climate Change. Geneva, Switzerland: IPCC; 2014

[2] NASA. The Effects of Climate Change. Available from: https://climate.nasa.gov/effects/. [Accessed: December 25, 2021]

[3] Ritchie H, Roser M. Urbanization: Our World in Data. 2018. Available from: https://ourworldindata.org/urbanization

[4] Wang J, Zhou W, Pickett STA, Yu W, Li W. A multiscale analysis of urbanization effects on ecosystem services supply in an urban megaregion. Science of the Total Environment. 2019;**662**:824-833

[5] Toparlar Y, Blocken B, Maiheu B, van Heijst GJF. A review on the CFD analysis of urban microclimate. Renewable and Sustainable Energy Reviews. 2017;**80**:1613-1640

[6] Furer F. Microclimates: What Are they and Why Do we Have Them? spectrumnews1. 2021. Available from: https://spectrumnews1.com/ca/la-west/weather/2020/12/30/microclimates--what-they-are-and-why-we-have-them-. [Accessed: December 25 , 2021]

[7] Liu B, Lian Z, Robert Brown D. Effect of landscape microclimates on thermal comfort and physiological well-being. Sustainability. Climate Change, Landscape in Transition and Tools for Sustainable Landscape Planning. 2019;**11**(19):5387

[8] Palme M, Salvati A. Urban Microclimate Modelling for Comfort and Energy Studies. 1st ed. Springer, Cham: Switzerland; 2021. p. 2021

[9] Geiger R, Aron RH, Todhunter P. The Climate near the Ground. 5th ed. Germany: Vieweg+Teubner Verlag; 1995. p. 1995

[10] Chen J, Saunders SC, Crow TR, Naiman RJ, Brosofske KD, Mroz GD, et al. Microclimate in Forest ecosystem and landscape ecology: Variations in local climate can be used to monitor and compare the effects of different management regimes. Bioscience. 1999;**49**(4):288-297

[11] Perry DA, Oren R, Hart SC. Forest Ecosystems. 2nd ed. Baltimore, Maryland, USA: The Johns Hopkins University Press; 2008. p. 2008

[12] United Nations Department of Economic and Social Affairs. 2018 Revision of World Urbanization Prospects. 2018. Available from: https://www.un.org/development/desa/publications/2018-revision-of-world-urbanization-prospects.html. [Accessed: December 25, 2021]

[13] Statista 2021. Share of Urban Population Worldwide in 2021, by Continent. 2021. Available from: https://www.statista.com/statistics/270860/urbanization-by-continent/. [Accessed: December 25, 2021]

[14] Sanusi R, Bidin S. Re-naturing cities: Impact of microclimate, human thermal comfort and recreational participation. In: Filho WL, Nagy JG, Borha M, Muñoz PDC, Magnuszewski A, editors. Climate Change, Hazards and Adaptation Options. 1st ed. Vol. 1. Switzerland: Springer, Cham; 2020. pp. 545-562

[15] de Souza DO, dos Santos Alvalá RC, Guedes M d N. Urbanization effects on the microclimate of Manaus: A modeling study. Atmospheric Research. 2016;**167**:237-248

[16] Xiong J, Lian Z, Zhou X, You J, Lin Y. Effects of temperature steps on human health and thermal comfort. Building and Environment. 2015;**94**(Part 1):144-154

[17] Izard CE. Emotion theory and research: Highlights, unanswered questions, and emerging issues. Annual Review of Psychology. 2009;**60**:1-25

[18] Wisconsin Department of Health Services. Carbon Dioxide. 2021. Available from: https://www.dhs.wisconsin.gov/chemical/carbondioxide.htm#:~:text=The%20levels%20of%20CO2%20in,of%20drowsiness%20and%20poor%20air. [Accessed: December 25, 2021]

[19] Climate Data Library. Q&a Basic Information - What Is the Average Air Pressure at the Surface of the Earth? 2021. Available from: https://iridl.ldeo.columbia.edu/dochelp/QA/Basic/atmos_press.html. [Accessed: December 25, 2021]

[20] Luo EK. Healthline. Hot and Cold: Extreme Temperature Safety. 2018. Available from: https://www.healthline.com/health/extreme-temperature-safety [Accessed: December 25, 2021]

[21] Kessel L, Johnson L, Arvidsson H, Larsen M. The relationship between body and ambient temperature and corneal temperature. AVRO. 2010;**51**(12):6593-6597

[22] Che Muhamed AM, Atkins K, Stannard SR, Mündel T, Thompson MW. The effects of a systematic increase in relative humidity on thermoregulatory and circulatory responses during prolonged running exercise in the heat. Temperature. 2016;**3**:455-464

[23] Randall DJ, Shelton G. The effects of changes in environmental gas concentrations on the breathing and heart rate of a teleost fish. Comparative Biochemistry and Physiology. 1963;**9**(3):229-239

[24] Verberkmoes NJ, Soliman Hamad MA, ter Woorst JF, Tan MESH, Peels CH, van Straten AHM. Impact of temperature and atmospheric pressure on the incidence of major acute cardiovascular events. Netherlands Heart Journal. 2012;**20**:193-196

[25] Pope CA, Verrier RL, Lovett EG, Larson AC, Raizenne ME, Kanner RE, et al. Heart rate variability associated with particulate air pollution. American Heart Journal 1999. 1990;**138**(5):890-899

[26] Kraus U, Schneider A, Breitner S, Hampel R, Rückel R, Pitz M, et al. Individual daytime noise exposure during routine activities and heart rate variability in adults: A repeated measures study. Environmental Health Perspectives. 2013;**121**(5):607-612

[27] Ahmed M. Low Heart Rate – Bradycardia Explained by a Heart Doctor. 2017. Available from: https://myheart.net/articles/low-heart-rate-bradycardia-explained-by-a-heart-doctor/. [Accessed: December 25, 2021]

[28] Pöyhönen M, Syväoja S, Hartikainen J, Ruokonen E, Takala J. The effect of carbon dioxide, respiratory rate and tidal volume on human heart rate variability. Acta Anaesthesiologica Scandinavica 2004. 2004;**48**(1):93-101

[29] Mordukhovich I, Coull B, Kloog I, Koutrakis P, Vokonas P, Schwartz J. Exposure to sub-chronic and long-term particulate air pollution and heart rate

variability in an elderly cohort: The normative aging study. Environmental Health. 2015;**14**:6-8. Article 87

[30] Huang J, Deng F, Wu S, Lu H, Hao Y, Guo X. The impacts of short-term exposure to noise and traffic-related air pollution on heart rate variability in young healthy adults. Journal of Exposure Science & Environmental Epidemiology. 2013;**23**:559-564

[31] Mullins JT, White C. Temperature and mental health: Evidence from the spectrum of mental health outcomes. Journal of Health Economics. 2019;**68**:3

[32] Sygna K, Aasvang GM, Aamodt G, Oftedal B, Krog NH. Road traffic noise, sleep and mental health. Environ-mental Research. 2014;**131**:17-24

[33] Kajtár L, Herczeg L, Hrustinszky T, Láng E, Bánhidi L. Influence of carbon-dioxide pollutant on human well-being and work intensity. In: HB 2006 - Healthy Buildings: Creating a Healthy Indoor Environment for People, Proceedings 2006. Lisbon: International Energy Agency; 2006

Chapter 9

Biology of Creativity: A Nondesign-Inspired Model to Enhance Creativity and Innovation Skills

Antonio Cruz Rodrigues, Cristiano Pedroso-Roussado, Joao Cunha and Ana Jorge

Abstract

From the merge of arts and crafts towards practice for mass production of desirability, consumption, and product development in a capitalist economy of scale, design has lost its natural ability of problem comprehension and tension alleviation. The modern world needs creativity, flexibility, and responsiveness embedded into design practices, mostly when a behavioural change, either individual or organisational, is intended. Still, the informality nature of the field is creating a gap between the study, research, and industrial design practice. Here, it is presented the Biology of Creativity Model (BoC), which is a design-by-analogy method that promotes an empowered design and creative practice through analysis of mostly biology reference texts for enhanced creative performance in a diverse array of contexts.

Keywords: design education, creativity, design model(s), analogical reasoning, bio-inspiration

1. Introduction

Design has been exploring artistic fields; however, its practice has been neglecting other fields of knowledge. The multifactor nature of current challenges is paving the way for wider perspectives about the practice of design [1]. Therefore, in recent years, biology has entered the design field as a discipline of inspiration and manufacture [2–4], through analogy [5–7] and abstraction-based design models and processes [8–10]. However, in contrast with other sciences, the design field is coined with a fragmentation of its body of knowledge, which makes it harder to solidify new discoveries into a standardised practice [1, 11]. Therefore, it is paramount to foster a simple language upon which further methodologies can be built on, increasing the design-led practitioners to solve planetary challenges in a more

IntechOpen

comprehensive way. In terms of creativity, the practice needs a build-up and a more concrete approach that could deviate the field from the elite, highly qualified top-tier ranks of organisations [12].

2. The problem of design

The current challenges for designers are systemic since the boundaries between objects, structures, systems, and processes are less defined and much more continuous and evolving [11]. It may seem obvious, but the speed of change unveils a challenging question for the current design paradigm, which is the inability of approaches developed in the twentieth century to solve twenty-first century challenges. We live in the storm of the fourth industrial revolution, and it is disrupting the production, management, and government realities we have known for several decades [13]. Therefore, designers need to be empowered by new competences in the physical, digital, and biological spheres [14].

2.1 A new educational approach for design and creativity

A novel transformation in design and creativity is necessary, one that encompasses a refocus of the design and creative practice, that merges the individual, industrial, and governmental behaviours that allow a healthier planet, safe from disastrous human activities that could potentially eliminate the risk of a global emergency. So, a suitable approach in design and creativity education must embrace a deep awareness for a competency-based model that paints a picture for the student rather than starting from the 'blank' [15]. If not diverse and independent, design and creativity education will form students with tendencies for those biased and standard practices and strategies, hampering their career development. Whilst seeking confirmation and comfort design students will undermine their full and unique potential [15–20].

2.2 A revitalised creative practice

At today's perspective, the practice of creativity has lost the critical thinking referential through which designers and other professionals ask themselves what the necessary 'whys' and 'hows' are for a good design and for a certain desired effect [21, 22]. The current limited scope of creative practice is seen in the absence of designers and artists in the decision-making I, not to mention the difficulties to standardise the formal methodological body related to ambiguity, inspiration, insight, *etc.* [22]. This is contrary to science or management areas, where there is abundant standardisation in the methodological frameworks [1, 11]. According to Whitney and Nogueira [1], 'design continues to produce novel solutions to specific problems but falls short in building knowledge that acts as a context for the various activities we call design'. Therefore, it is urgent a novel and rigorous approach to empower designers and creative professionals. Simultaneously, it is also urgent to exponentially increase the creatives' comprehension of the broader setting where they operate. Additionally, the focus of comprehension in creativity has been too linked to the creation, leaving knowledge gaps in the creative process. The creative process is a dynamic set of acts that deserves better attention and study to depart from the incompleteness that the theories of creativity proposed so far [23, 24].

3. Biology of creativity model

3.1 Definition and purpose

The Biology of Creativity (BoC) model is an internationally acclaimed process of innovation and creativity that makes use of reference texts originated essentially from biology to perform critical analogic thinking. This model has been implemented and tested worldwide through a set of workshops in various contexts, such as educational and corporate programs. The underlined provocation follows the question: What we ought to know about design if we only know and stay within the design field?

Therefore, the purpose of the BoC model is to facilitate the relationships between different areas of knowledge by the meaning of straightforward abstractions that will interact among themselves, creating a fertile ground for creativity and innovation to happen in a multitude of contexts [25]. This way the BoC model produces abstractions that allow the better comprehension of dissimilar phenomena. The resulting descriptive equation occurs from their detected similarities. The ultimate goal is to empower the creative and innovative processes of people and organisations, irrespective of their *modus operandi*, allowing a healthier and necessary maturation period in order to achieve a more complete productivity.

3.2 The BoC model process

One word is worth more than a thousand images. The interaction between texts is a relevant form of comprehension of psychological, social, and economic phenomena [26, 27]. Each text, paragraph, word, and each element gathered from a group of texts form a coherent content that originates a collective and multiple entity that evolves and renews itself into new, different, and unexpected perspective-dependent configurations. Therefore, the comprehension of texts is a powerful and simple method to create an ever-renovated movement of comprehension.

In physics, anisotropy is usually defined by a body's responsive behaviour when stimulated to act in different spatial directions [28, 29]. Thus, the same can be assumed when an array of discrepant perspectives (that may be related to modern trends, music, arts, or even to scientific fields) are applied to the comprehension of texts. This event creates an uncomfortable perception that generates curiosity, reflection and, potentially, innovation and creativity.

The incorporation of a single new element creates a disequilibrium due to the non-stable character of the properties that are originated from the comprehension of a group of texts. Such disequilibrium is not trivial and encompasses the basic notion of resiliency happening throughout the learning process, which may be defined by the interchange of equilibrium and disequilibrium states [28, 30].

According to literature, there are several benefits of using design-by-analogy models. They are mostly related to the novelty of ideas, and they were defined as the 'threshold of open goals' (keeping the original goals in the short-term memory while performing analogical thinking), 'narrowed scope of design space' (through the design fixation problem), 'full functionality points to (holistic) system level of design, limited functionality points to limited scope of problem' (by addressing the big picture of the design problem), 'difficulty in mapping narrow functionality' (allowing a better comprehension through the use of less functional analogies), 'more functionality equals to more opportunity for analogical transfer', and 'potential sweeter spot in level of functionality' (through the combinatorial effect of the analyses) [31]. Although, these assumptions were not directly addressed in the BoC model. The intention of the BoC model was to simplify the analogical effort of the participants by making use of a stepwise process with facilitating tools to expand and improve the fluidity of creativity and innovation processes, taking advantage of reference texts mainly sourced from the biological sciences.

3.3 Methodology

The BoC's methodology starts from the assumption that all companies, organisations, processes, products, territories, *etc.*, continuously compare past experiences, and design and update their point of view as active members of society [32–35].

In a broader design of innovative and creative content and processes, a relationship between disperse elements is established which forces the design approach to respond to the new information. In summary, relationships might occur in two main spectra: spatially—resulting from distant elements—and temporally—resulting from time—separated elements. Therefore, these two spectra potentially originate constructs that can be homogeneous (constituted by similar elements) or heterogenous (constituted by dissimilar elements).

In practice, the BoC model is constituted by four phases and two facilitating working tools. Thus, the BoC model is a closed model represented by a guidance reference letter that is explained as follows:

i. Phase 1: Information is characterised by a contextualisation of the texts to be analysed;

ii. Phase 2: Objectives are characterised by the definition of what is meant to be achieved;

iii. Phase 3: Actions are characterised by the enunciation of the planned activities;

iv. Phase 4: Project is characterised by the practical definition of the project structure and consequent action plan.

3.4 Scientific assumptions

The scientific conjecture of the three first phases—information, objectives, and actions—is developed by means of reflection exercises based on analogic thinking. Every moment and thinking step of the BoC model are based on the exposition of reference texts in the form of paragraphs. These paragraphs are curated from a diverse array of fields of knowledge, predominantly from biology, but also from economy, anthropology, archaeology, genetics, psychology, sociology, cosmology, history, *etc*. The intentional appropriation of other disciplines has the intention to provoke associations between the texts, images, and subsequent constructs. The BoC model promotes analogical thinking between these texts and organisations, companies, products and services, territories, *etc*. Moreover, the achieved text-text or text-image-construct associations resemble a spatial resonance—possessing a semantic elasticity through approximation and separation—, and a temporal resonance—in a chronological perspective.

The correspondent extrapolation to putative new realitIes is participant-dependent since they belong to different contexts (*e.g.*, companies, organisations, processes, products, territories, *etc.*). In each of the first three phases, it is asked three questions mainly coming from the biology field. These questions invite the participants to further ask questions in relation to their own setting, in an innovation and creative manner. Phase 4—project is developed in terms of a new project, taking into consideration its characterised elements and their relation to the theoretical scope of the previous Phases.

3.5 Phases: information, objectives, actions, and project

3.5.1 Phase 1: information

During this phase the goal is to answer the following questions through the reference texts as guidance:

A—Are we homogenising the environment?;

'Among other organisms, the genetic variability in insects is negatively affected in more populated human area. Additionally, the human interference in the more remote areas is stressing insects' populations as well. As an example, researchers have found that the introduction of European plant species in Antarctic remote islands decreased the local insect's species' richness. We are homogenising the environment' [36].

A—Are we homogenising the products?;

Analogically, we may state that the variability between products is less in more populated areas. Additionally, the interference that exotic products exert on local cultures is contributing to the local products' extinction, or at least overshadowing them. We may say that the introduction of exotic products in more remote locations tend to uniformise local products. Therefore, we are homogenising the products.

B—City birds have an increased capacity to approach strange objects?

'The neophiliac attitude, *i.e.*, the animals' ability to approach and interact with uncommon objects, is seen for instance in the increased capacity that city birds have to approach strange objects in comparison to non-city birds. Thus, the city birds are less cautious than usual when facing the new and strange objects present in their environment for evolutive reasons' [38].

B—What companies, organisms, *etc.*, have the capacity to approach uncommon environments?

Analogically, we may state that there are neophiliac companies and enterprises, with the ability to approach and explore uncommon environments. This behaviour manifests itself in the ability and interest that multinational companies have in approaching strange contexts in a swift manner compared to local ones. Multinational companies are less cautious than usual in relation to the risk they assume for new and unknown objects present in their environment.

C—Are we becoming only capable of prosper in specific habitats?

'The evidence leads us to believe that the human species—generalist and able to colonise every environment—is quickly becoming in a specialist species, only able to prosper within specific habitats, like our cities. We can describe our cities using the concept of ecological niche, embodied by the habitat's essential characteristics; the only place where our specialisation path is allowing the best survival chances' [39].

C—What companies, organisms, *etc.*, have the capacity to approach uncommon environments?

Analogically, we state that companies, organisms, *etc*., possessing a generalist profile (able to adapt different environments) are becoming specialists, only able to prosper in specific habitats/contexts/markets. We may also describe most companies using the concept of ecological niche, embodied by the habitat's essential characteristics; the only place where our specialisation path is allowing the best survival chances.

After this phase, it is introduced a theme associated with one sociological context that aims to:

i. Its own relation to the three analysed reference texts;

ii. Its succession (by concatenation) outcome, seen as cause-effect relation;

iii. The generation of emergent properties, where each reference text's limits are blurred allowing the appearance of culturally enriched undefined constructs.

3.6 Phase 2: objectives

During this phase the goal is to answer the following questions through the reference texts as guidance:

A – Do the communities only last while persist the conditions that helped them to start?

"Groups of species temporally and spatially bounded create the illusion of stability, but these communities can only last while the initial base conditions persist. When the biome conditions change, it does not matter if temperature, acidity, seasonality, or rainfall, many present species loose equilibrium. To some that means migration, following the environmental shift throughout the landscape like many plants did during the last glacial period" [37].

A – Do the communities of companies, organisms, ... and products, can only last while persist the conditions that helped them to start?

Analogically, communities of companies, organisms, ... and products, in a temporal and spatial symbiotic relationship, may create the illusion of stability, because in practice they can only last while the initial base conditions persist. So, when the conditions change, part of their elements may lose balance. To some companies, organisms, ... and products, this condition may mean migration to other markets.

B—Do animals and plants have a kind of evolutive capital?

'The permanent genetic variation is, thus, some sort of evolutive capital, that allows the species in possession of such capital to take advantage of its genetic wealth, and to quickly produce any combination of genes required by an altered environment. Animals and plants, for instance, in need to adapt to an added element in the environment, do not need to wait to mutations to happen. In most cases, the necessary genetic variations are already there' [38].

B—Do companies, organisms, *etc*., and products have some sort of evolutive capital?

Analogically, we may observe that companies, organisms, *etc*., and products in possession of evolutive capital—allowing them to take advantage of their genetic wealth, such as experiences, technologies, and processes—quickly produces any combination of characteristics required by an altered market environment. Companies, organisms, *etc*., and products that, for instance, need to adapt to an added market element, do not need to wait until deep mutations occur. In most cases, the necessary variations are already there.

C—Can we have different results when starting from the same DNA?

'We know that the human skin can darken tan if exposed to sunbathe and the hair can lighten. In most of our physical features both nature and environment's influence play a role. And it applies to all animals. That is what is called plasticity, in which different results may occur when starting with the same DNA. For instance, when we observe a difference in the colour between the populations of the same animal or plant species living in different ecosystems, it may be an example of plasticity' [38].

C—Can the same inherent characteristics in companies, organisms, *etc*., and products, may have different results?

Analogically, in some companies, organisms, *etc*., and their products, the inherent characteristics are not everything. There is a group of emergent characteristics that result when the environment's influence enters into play. That is what is called plasticity, in which different results may occur when starting with the same inherent characteristics. For instance, when we observe differences in different environments, it may be an example of plasticity.

After this phase, similarly to the previous phase—information, it is introduced a theme associated with one sociological context that aims to:

i. Its own relation to the three analysed reference texts;

ii. Its succession (by concatenation) outcome is seen as cause-effect relation;

iii. The generation of emergent properties, where each reference text's limits are blurred allows the appearance of culturally enriched undefined constructs.

3.7 Phase 3: actions

During this phase the goal is to answer the following questions through the reference texts as guidance:

A—Are there organisms that adapt to other organisms that in turn can evolve?

'There is one type of evolution, empowered by the unlimited possibility of the adaptation process, that is observed when one organism must adapt not to something physical, but to other organism, that in turn may evolve. The evolution of one part stimulates the evolution of the other and has one ecological interaction that connects both organisms as a net result, as two engaged species in a single continuous relationship of predator/prey, evolving continuously, both in the scope of processes, or the offensive/defensive techniques' [38].

A—Are there companies, organisms, *etc*., and products that adapt to other entities that in turn can evolve?

Analogically, there is one type of evolution in companies, organisms, *etc*., and products that is empowered by the unlimited possibility of the adaptation process, that is observed when one company, organism, *etc*., or a product evolve and adapts to other company, organism, *etc*., or product, that in turn can evolve. The evolution of one part stimulates the evolution of the other and has one interaction that connects companies, organisms, *etc*., and products between themselves, becoming engaged in a continuous relationship with the purpose of, for instance, one overpowering the other, or that they progress together.

B—Can two plants unite into one unique individual under particular circumstances?

'By Merging two bionts—independent living material units—through a graft, we may in theory, create a new plant that unifies the positive characteristics, such as two apple trees, *i.e.*, a new apple tree that is simultaneously resistant to drought and able to produce magnificent apples. Two similar plants that face permanent contact, for instance, at the trunk or between branches, can, under particular conditions, merge into a unique individual, by inserting a part into the other' [39].

B—Can two companies or products unite into one unique individual?

By merging two companies, organisms, *etc.*, or products through a graft, we may in theory, create a new company or product that unifies the positive characteristics, like two products, *that is*, a new product that is simultaneously adaptable to two different environments. Two companies, organisms, *etc.*, or products that face permanent contact, for instance, by being present in the same market, can, under particular conditions, merge into a unique company, organism, *etc.*, or product, by inserting some characteristics from one into another.

C—Is it possible that some temporary combinations last longer than others?

'There is no differential reproduction in the abiotic world, but we may get variations of differential persistence: some temporary pieces' combinations last longer than others, thus having more time for reviewing and adjustments. The differential persistence must, somehow, transform itself gradually into differential reproduction, multiplying its advantage by... multiplication. The differential reproduction may happen because of a "lucky collision" of two non-related elements, "serendipity", with a happy outcome, as opposition to a "demolition" with a destructive outcome' [40].

C—Is it possible that some temporary combinations of elements last longer than others?

Analogically, both in the context of companies, organisms, *etc.*, as in products, some temporary combinations of elements, characteristics, technologies, materials, markets, *etc.*, last longer than others, thus having more time for reviewing and adjustments. The differential persistence must, somehow, transform itself gradually into differential reproduction, multiplying its advantage by... multiplication. The differential reproduction may happen because of a 'lucky collision' of two non-related elements, 'serendipity', with a happy outcome, as opposition to a 'demolition' with a destructive outcome.

After this phase, similarly to the previous phases—information and objectives, it is introduced a theme associated with one sociological context that aims to:

i. Its own relation to the three analysed reference texts;

ii. Its succession (by concatenation) outcome is seen as a cause-effect relation;

iii. The generation of emergent properties, where each reference text's limits are blurred allows the appearance of culturally enriched undefined constructs.

3.8 Phase 4: project

The three questions presented during this phase show the idea of concatenation as an assumption, which results from a previous association, and in succession with the three reference texts discussed during the previous phases. Thus, starting from Phase 1—information, until Phase 3—actions, there is an objective rationale directed to the context of project, and under the scope of a cause-effect relationship.

A—What is the reference board, the context, and the elements of the project?

This happens when each project's proposed element has a meaning that depends upon a reference context.

B—What are the combinatory probabilities/possibilities between the project's elements?

This happens because of the possible number of modes, in which a group of elements, context, and reference board, can be exposed or combined, thus resulting in a possible project's 'object'.

C—What are the emerging properties resulting from the pertinent combinations?

This happens when the project's 'object' results in a property/characteristic that does not come from the description of its individual elements but from the interaction and junction between the elements to form an entity.

During development of Phase 4—project, it is introduced themes from the scope of sociology, anthropology, culture studies, *etc.*, that stimulate and focus a better and broader contextualisation for the working projects, that polarises themselves among a strategic dimension like the incorporation of design within organisations; passing through communication actions like the creation of narratives; until the corporeality of 'products'.

4. Extenders and Unlockers

Extenders and Unlockers are active working tools that are used during the entire BoC process. Extenders have the objective of extending a particular subject or text towards another reference resource, or towards a similar theme, thus improving its comprehension. They are usually used during Phase 1—information, Phase 2—objectives, and Phase 3—actions. Unlockers are mostly used during the Phase 4—project, and their objective is to unlock certain project moments that, for some reason, presented hurdles for the creative progression. The Unlockers are texts that allow a better decision-making process based on the elements of the project and let the widening of the observing and thinking ability of the audience. Therefore, Unlockers present a practical feature of the project, improving the deliberation of its actionable parameters.

In summary, Extenders and Unlockers' aim is to facilitate the learning of concepts and the materialisation of project's constructs. These two tools are used in accordance with particularities related to the target audience, the quality, and characteristics of the previous considerations, analogies, and associated themes analysed and generated.

4.1 Examples of extenders

As an example, we may refer to a first text that can be introduced in the reflection process, that points to the concept of 'entanglement' that results when quantum particles entangle each other and thereafter cannot be described as independent particles. This concept allows for instance to reinforce texts and questions enunciated in the Phase 3—actions, such as 'Can two plants unite into one unique individual under particular circumstances?'; which in turn refers to an associated question: 'Can two companies or products unite into one unique individual?'

As a second example of Extenders, we may refer to a second text that can be further introduced which mentions to the idea of 'speciation' that results from the process upon which natural populations evolve to become distinct species, generating variety. Similarly, to the previous extender text, it allows for instance the reinforcement of texts and questions enunciated during the Phase 2—objectives, such as 'Can the same inherent characteristics in companies, organisms, *etc.*, and products, may have different results?'

4.2 Examples of Unlockers

As examples of Unlockers, we may refer a first text that points to the concept of 'counter examples', which results from the association between negative and positive examples, which in turn focus on the relation between characteristics that oppose each other and so they can be contained or excluded into a project's 'object', allowing a better objects' definition.

A second example, acting in sequence from the previous Unlocker, refers to the concept of 'polarity', which results from the higher or lower tension between two opposing elements, and that are defined in the scope of the same project's 'object', or between 'objects'.

5. Reflection's example

Following the next examples, the purpose of the BoC model can be stated in thesis as the reflection and identification of elements aiming at the careful definition of an action plan.

5.1 Example 1

Let us take as example one real 'problem' that can interfere in the context of companies', organisations', processes', products', territories', *etc.*, activities, that is mentioned in the Jeremy Rifkin's book entitled 'The Zero Marginal Cost Society: The Internet of Things, the Collaborative Commons, and the Eclipse of Capitalism':

'Everywhere a younger generation is sharing bicycles, cars, houses, clothes, and many other numerous articles, preferring the access to the possession. An increasing number of young adults choose not to have branded articles in favour of more generic ones or brands with a higher purpose. These youngsters are revealing themselves as much more interested in the materials' goods usability value, rather than in the interchangeability or status value. An economy of share comprised of collaborative prosumers is, in its genesis, a more empathetic and less materialistic one. The unbridled consumption was replaced by an economy of sharing' [41].

This reference text refers to the essential question—Are companies, organisations, processes, products, territories, *etc.*, prepared for a shift in the economy that tends to a greater extent to sharing and to the common goods? In this context, what elements are necessary to answer this new reality, in the scope of the questions enunciated in the BoC model, such as 'What companies, organisms, *etc.*, have the capacity to approach uncommon environments?', or, 'Are companies, organisms, *etc.*, only capable of survive in specific markets?' In summary, do companies, organisms, *etc.*, that show elements pre-adapted to the new context have more chances to adapt and thrive in the new reality characterised by the change in the economic paradigm?

5.2 Example 2

Let us take as example one real 'problem' that can interfere in the context of companies', organisations', processes', products', territories', *etc.*, activities, that is mentioned in the Gilles Lipovestky's book entitled 'Le Sacre de L'autenticité':

'There is no exclusive value attributed to the singularity of subjects, but also to the singularity of objects; there is no exclusive fidelity's glorification to the subjective

DOI: http://dx.doi.org/10.5772/intechopen.112738

self, but also to the fidelity of brands in relation to themselves; there is no exclusive appreciation for the 'natural' behaviour of people, but for the ecological products that respect the environment as well. The authenticity ideal, originally intra-personal, penetrated in the universe of the 'things' and of the company: we desire, everywhere, meaning, truth, transparency, naturality, sincerity, and fidelity to the new self. The hypermodern universe is characterised by the authenticity ethics' extension to the products' saleable sphere' [42].

This reference text refers to the essential question—Are companies, organisations, processes, products, territories, brands, *etc.*, prepared for a fidelity ideal towards themselves, one of authenticity that penetrates the universe of products, and to the creation of transparent and truthful narratives? In this context, what elements are necessary to answer this new reality, in the scope of the questions enunciated in the BoC model, such as 'Are we homogenising the products?', and 'Do companies, organisms, *etc.*, and products have some sort of evolutive capital?' In summary, do companies, organisms, brands, *etc.*, have an evolutive capital that distinguish themselves as the singular authentic subjects, that allow them the creation of transparent and truthful narratives? And that in a way that allows them the creation of non-homogeneous products in the cultural context that tends to the globalisation trend and loss of identity?

6. Evaluation and results parametrisation

The BoC model, its process and setting, throughout its four phases—information, objectives, actions, and project, permit the identification of a transversal structure of innovation and creativity, and allows the evaluation and parametrisation of the obtained results. Hence, by the identification of variables, features, and patterns, the participants improve their operative lexicon, thus granting the development of the BoC model itself as well.

Specifically, the BoC model makes use of four expansive parameters for its evaluation and parametrisation:

i. Creative cycle;

ii. Imaginative capacity;

iii. Patterns' identification; and

iv. Lexicon enrichment.

6.1 Creative cycle

The first parameter is related to the creative cycle offered by the BoC model and it can be defined by four distinct creative moments observed by the:

i. Capacity of identifying pertinent texts;

ii. Capacity of apprehend and formulate abstractions; and by the

iii. Capacity in completing the creative working sequence, and generating new objects of knowledge.

6.2 Imaginative capacity

The second parameter is related to the imaginative capacity in analysing dissimilar texts with a variable degree of distance, which implicates a variable and associated imaginative capacity to build interesting, innovative, and creative narratives.

6.3 Identification of patterns

The third parameter is related to the identification of patterns. It results from the identification of words and meanings from the exposed texts that motivated analogic thinking and further lead to a better comprehension of practical realities. This capacity depends on the audience and its efficacy varies depending on the tailored introduction of different perspectives.

6.4 Lexicon enrichment

The fourth and last parameter is related to the lexicon enrichment throughout the process, and it is observed by the increase of the available lexicon. In practice, the quality of the available lexicon is observed by the participants' improvement of thinking, comprehension, and description of the exposed realities. The Phase 4—project is the one that enables the validation of the achieved progress, by the evaluation of more elaborated and meticulous definitions.

7. Application

The BoC model here presented was previously developed, tested, and optimised in a set of workshops realised in the context of University Programs (Bachelors, Masters, and PhDs programs) and Research Groups, Business Associations, and blended groups of students and professionals, in Spain, Italy, Brazil, Portugal, Japan, and China.

The BoC model can be applied during workshop sessions that are developed during an ongoing methodology of 16 hours. The practice is preferably divided into four sessions occurring on different days. A single phase—information, objectives, actions, and project—is developed during each session.

At the beginning and at the end of each workshop session and in accordance with the audience's profile, four questions are asked to implicitly influence every BoC model phase—information, objectives, actions, and project. They have the goal to analyse the progression and improvement in the reflection and comprehension, and innovation and creativity skills of the participants during the entire process. Therefore, the BoC model assumes the main purpose of improving such skills in a tangible way. The questions are:

i. What is the surrounding sociological perspective of your organisation?

ii. What are the actionable points that you wish to obtain and develop?

iii. What 'products' do you hope to make?

iv. What are the particular features that these new 'products' must possess?

DOI: http://dx.doi.org/10.5772/intechopen.112738

Consequently, the answers to the same questions happening at the end of the workshop must reveal an improved capacity of creative competence, observed by a more complete 'Creative Cycle', a greater 'Imaginative Capacity', a better 'Identification of Patterns', and a clear 'Lexicon Enrichment'.

Additionally, these four evaluation parameters serve the diagnostical purpose for a variety of audiences, such as companies and organisations, and to creative practitioners (such as designers) as well. These parameters allow for the tangible evaluation of competences for innovation and creativity. So, it is possible that one entity may be able to critically observe and comprehend practical realities using abstractions, although being unable to complete the creative process and consequently unable to generate new objects of knowledge. This setting is casually observed, and it may be explained by the homogenisation of the technical and managerial competences within the organisation. Other causes may also explain this dogmatic setting like the absence of progressive learning practices [32], the exclusive orientation towards specific and known markets [43, 44], thus avoiding the risk-taking activities associated with the exploration of unfamiliar environments [45, 46]. Such limitations are aggravated by the utilisation of a poor lexicon in the definition of strategies, objectives, and other leadership decisions [22, 24, 47, 48].

8. Conclusions and future work

In the case that companies, organisations, processes, products, territories, *etc.*, in a first moment must have the capacity of moving between apparently dispersed signals, in an expanding attention, that widen the horizon, in a second moment, they must have the capacity of concentrating themselves, ignoring and selecting the pertinent signals. It is this transition process, from a space of attention to a space of concentration that characterises the BoC model, throughout the four phases—information, objectives, actions, and project. Additionally, it is also the main objective in the context of companies, organisations, processes, products, territories, *etc.*, to provide them with a new biology of creativity and innovation, making them able to widen and coordinate multiple areas of knowledge targeting the broad scope of the creative practice. And to provide them the capacity to reconfigure, optimise, and continuously update their activity in a broader perspective of content and processes' design, but mostly in a more active perspective towards a richer, more complex, and challenging society.

Lastly, and referring to the initial question—in the context of a 'Biology of Creativity', is there a maturation time necessary for a company, organism, *etc.*, or product to reach their complete productivity?—Yes, as an embryo develops until it reaches adulthood, there is the need to some years for a child to reach her complete productivity, fully able to be integrated in society. Analogically, companies, organisations, processes, products, territories, *etc.*, also need time to make reviews and adjustments to fit into society.

Contrary to other disciplines where the focus is dedicated to the understanding and analysis of the target phenomena, design, and creativity are disciplines focused on the synthesis and creation of novel things [11, 24]. Hence, it is obvious that design and creative teaching and practice must incorporate a wide range of disciplines, mostly outside of the design realm, such as business administration, ethics, and sociology, biology, and ecology, to be able to better fit into a multisectoral context [24]. Accordingly, the noticeable climate crisis constitutes evidence of the opposite point, that design has promoted an unbalanced practice in a world hit of constant crises.

The power of the BoC model relies on the use of non-design text references to expand the comprehension, lexicon, and action of the participants, irrespective of their provenience. The observed creativity empowerment is a natural achievement of the BoC model that is encouraged by the reasonable easiness of interpretation of the novel texts, allowing a practical comparison with the current reality of each participant's sociological context. Thus, the participant is urged to act upon the new ability to analyse and compare, creating space for a more motivated, active, and responsible individual in an ambitious and empowered action plan.

The BoC model forces the questioning of diverse elements and features that constitute the structure of a manifold of cases studies. Thus, the BoC model helps creativity professionals, innovators, companies and organisations, and other entities in general, to develop capacities of reflection and action in the innovation and creativity setting. This is achieved by employing an easy-to-apply stepwise framework constitute of different phases that walk the talk of creativity. The discomfort some participants feel while working with the BoC model is normal and should be interpreted as natural. However, those discomforts must be attenuated during the work promoted by the BoC model.

In terms of the scientific literature associated with the BoC model, design-by-analogy methods are what are explored during the work plan. These methods mainly represent a way to solve design fixation [49]. Although fixation may be helpful for developing extremely focused projects and tailoring product development efforts for known markets, it is valuable to solve fixation to potentiate creativity and help guide further projects that relate to challenges coming from unpredictable events and for perhaps unknown markets. Therefore, further research must be performed in order to decipher the complex causes of design fixation and individual and collective limitations to creativity and innovation, such as cognitive entrenchment [50], confirmation bias [51], psychological ownership [52], and tolerance for ambiguity [53, 54].

Similar to what was observed by others [55], the BoC model also promotes the rise of ideas from the analogies permitted by the reference texts and throughout the different BoC's phases—information, objectives, actions, and projects. Additionally, the particularities of the participants, such as their expertise and career paths, enrich the learning experience and the constructs generated [33]. Therefore, the introduction of novel texts, perspectives, and other elements, into the innovation and creativity processes allows a satisfactory approximation to the real world [25].

Despite its international application and very positive appraisal, the BoC must be further developed and improved. So, the next steps may rely on the application of the BoC model to more audiences that could continue the formal validation of its assumptions, processes, and evaluation methods, and identify limitations that could be further explored. Moreover, the last phase—project, should be further developed in order to allow an upscaled approach for participants from high valued enterprises and governmental bodies, with a diverse background and expertise. However, it must be interesting to develop tailored project's sub frameworks that could encompass the manifold variables coming from different industries and sectors. Nevertheless, an improved validation must be addressed in upcoming research endeavours to enable standardisation of research and implementing practices of the BoC model [56]. Still, we believe that the BoC model presents an original, novel, proper, and detailed stepwise process that can be widely deployed and parametrised for self and autonomous use of individuals and organisations to promote a better practice regarding creativity.

A possible critique of the BoC model is the constant appropriation of references that are distant from the design and creative education and practice standards, increasing the chance of misinterpretation of seminal concepts and theories. Another critique is the fact that some of the referenced texts may be outdated, and novel perspectives are now more accepted in the scientific community. However, these two critiques do not hamper the strengths of the BoC model since its aim relies on the analogical power of the references to provide enhanced creativity and innovation skills [11, 47, 57]. Nevertheless, the scientific rigour of the BoC model is paramount and it must be pushed forward in the design and creative practice in general to strengthen the activity of creative professionals whatsoever, where a broader knowledge and experience are essential.

Acknowledgements

The authors, specifically ACR, want to thank the following people who facilitated the work here developed: Lorenzo Secco from Università Iuav di Venezia, Italy; Stefano Parissoto from Unindustria Servizi & Formazione Treviso Pordenone, Italy; Katia Gasparini from Accademia delle Belle Arti di Verona, Italy; Chiara Cibin from Academia delle Belle Arti di Bologna, Italy; Marinella Ferrara from Politecnico di Milano, Italy; Richard Perassi from Universidade Federal de Santa Catarina, Brazil; Marco Aurélio Lobo Junior from Centro Universitário IESB, Brasília, Brazil; João Luís Rieth from Universidade Extremo Sul Catarinense, Criciúma, Brazil; Thiago Meneghel Rodigues from Laboratório de Inovação em Desenvolvimento Regional Empreendedorismo – LIDERE, Lages, Brazil; and Pieter Sprangels from Karel de Grote University, Antwerp, Belgium.

In particular, ACR wants to acknowledge his inspiration for the development of this work. These people always left a motivational disquiet, promoting a critical view about 'things', in a permanent dynamic of added complexity. ACR had the honour of befriend all of them, sharing in the present, with those who are still among us, and in the past, with those who are not in the physical world anymore, but who are still present in his thought almost every day, an intellectual intimacy that increased the meaning of almost 'everything' he does. These people are great for their culture, generosity, and mostly for their contribution to the way we understand our surrounding 'realities', and for everything they add to our lives. They are: Di Salvatore, painter, sculptor, *etc.*, who taught ACR in the Scuola Politecnica di Design di Milano, and guided him through the importance of the Vision Science; Bruno Munari, designer, artist, *etc.*, who taught ACR in the Scuola Politecnica di Design di Milano, and guided him through the practice of the Analogical Thinking; Attilio Marcolli, architect, designer, *etc.*, who taught ACR in the Scuola Politecnica di Design di Milano, and guided him through the importance of Maths in Design; Gillo Dorfles, philosopher, art critic, *etc.*, who guided ACR through the Design Philosophy; Fernando Carvalho Rodrigues, scientist, physicist, *etc.*, who guided ACR through the importance of the Logical Thinking; Paulo Cadete Ferrão, engineer, explorer, *etc.*, who stirred ACR to the importance of sustainability in Innovation and Creativity; and lastly, Pietro Zennaro, philosopher, essayist, *etc.*, who encouraged ACR through the importance of the irreverence and questioning of 'life'.

Authors' contributions

ACR and JC conceptualise and validated the model. CP-R and AJ contributed with inputs to the model. ACR and CP-R wrote the original draft. All authors reviewed and edited the final draft of the manuscript.

Funding

This research did not receive any specific grant from funding agencies in the public, commercial, or not-for-profit sectors.

Declaration of interest statement

The authors declare that they have no known competing financial interests or personal relationships that could have appeared to influence the work reported in this paper.

Author details

Antonio Cruz Rodrigues[1], Cristiano Pedroso-Roussado[2,3*], Joao Cunha[1] and Ana Jorge[4]

1 Universidade Lusófona de Humanidades e Tecnologias, Lisbon, Portugal

2 ITI/LARSyS, Instituto Superior Tecnico, Universidade de Lisboa, Lisboa, Portugal

3 UNIDCOM/IADE, Lisbon, Portugal

4 NOVA LINCS, NOVA School of Science and Technology, NOVA University Lisbon, Portugal

*Address all correspondence to: cristiano.roussado@tecnico.ulisboa.pt

References

[1] Whitney P, Nogueira A. Cutting cubes out of fog: The whole view of design. She Ji: The Journal of Design, Economics, and Innovation. 2020;**6**(2):129-156

[2] Cao G, Sun Y, Tan R, Zhang J, Liu W. A function-oriented biologically analogical approach for constructing the design concept of smart product in industry 4.0. Advanced Engineering Informatics. 2021;**49**:101352

[3] Chirazi J, Wanieck K, Fayemi PE, Zollfrank C, Jacobs S. What do we learn from good practices of biologically inspired design in innovation? Applied Sciences. 2019;**9**(4):650

[4] Yargın GT, Firth RM, Crilly N. User requirements for analogical design support tools: Learning from practitioners of bio-inspired design. Design Studies. 2018;**58**:1-35

[5] Fu K, Moreno D, Yang M, Wood KL. Bio-inspired design: An overview investigating open questions from the broader field of design-by-analogy. Journal of Mechanical Design. 2014;**136**(11):111102

[6] Moreno DP, Blessing LT, Yang MC, Hernández AA, Wood KL. Overcoming design fixation: Design by analogy studies and nonintuitive findings. AI EDAM. 2016;**30**(2):185-199

[7] Nagel JK, Schmidt L, Born W. Establishing analogy categories for bio-inspired design. Designs. 2018;**2**(4):47

[8] Deng X, Wang HH, Liu CY, Wang YH. Can abstraction help ideation? A case study on biologically inspired design. International Journal of Technology and Design Education. 2022;**32**:2725-2747

[9] Goldschmidt G, Sever AL. Inspiring design ideas with texts. Design Studies. 2011;**32**(2):139-155

[10] Lenau TA, Keshwani S, Chakrabarti A, Ahmed-Kristensen S. Biocards and Level of Abstraction. Milano, Italy: International Conference on Engineering Design - ICED15; 27-30 July 2015

[11] Meyer MW, Norman D. Changing design education for the 21st century. She Ji: The Journal of Design, Economics, and Innovation. 2020;**6**(1):13-49

[12] Gormley K. Neoliberalism and the discursive construction of 'creativity'. Critical Studies in Education. 2020;**61**(3):313-328

[13] Schwab K. The Fourth Industrial Revolution. London, England: Currency; 2017

[14] Wilde D. Design research education and global concerns. She Ji: The Journal of Design, Economics, and Innovation. 2020;**6**(2):170-212

[15] Baha E, Koch M, Sturkenboom N, Price R, Snelders D. Why am I studying design? In: Boess S, Cheung M, Cain R, editors. Synergy - DRS International Conference; 2020. pp. 11-14

[16] Adams RS, Daly SR, Mann LM, Dall'Alba G. Being a professional: Three lenses into design thinking, acting, and being. Design Studies. 2011;**32**(6):588-607

[17] Gray C. Locating the emerging design identity of students through visual and textual reflection. In: Lim Y, Niedderer K, Redström J, Stolterman E, Valtonen A, editors. Design's Big Debates - DRS International Conference. Umeå, Sweden; 16-19 June 2014

[18] Kosonen K. Finding One's Own Way In Design-Reflections on Narrative Professional Identity. Aalto, FInland: Aalto University; 2018

[19] McDonnell J. Becoming a designer: Some contributions of design reviews. Analyzing Design Review Conversations. 2015:355-371

[20] Tracey MW, Hutchinson A. Reflection and professional identity development in design education. International Journal of Technology and Design Education. 2018;**28**(1):263-285

[21] Dorst K. Understanding Design: 150 Ways of Looking at Design. Amsterdam, Netherlands: BIS; 2003

[22] Oleynick VC, Thrash TM, LeFew MC, Moldovan EG, Kieffaber PD. The scientific study of inspiration in the creative process: Challenges and opportunities. Frontiers in Human Neuroscience. 2014;**8**:436

[23] Sternberg RJ, Karami S. An 8P theoretical framework for understanding creativity and theories of creativity. The Journal of Creative Behavior. 2022;**56**(1):55-78

[24] Walia C. A dynamic definition of creativity. Creativity Research Journal. 2019;**31**(3):237-247

[25] Graff D, Meslec N, Clark MA. Perceived analogical communication in design teams: Development and validation of a scale. Design Studies. 2020;**67**:55-76

[26] Kintsch W. The role of knowledge in discourse comprehension: A construction-integration model. Psychological Review. 1988;**95**(2):163

[27] LaRusso M, Kim HY, Selman R, Uccelli P, Dawson T, Jones S, et al. Contributions of academic language, perspective taking, and complex reasoning to deep reading comprehension. Journal of Research on Educational Effectiveness. 2016;**9**(2):201-222

[28] Parisi G. A Minha História na Física: O voo dos estorninhos e outros sistemas complexos. Lisboa, Portugal: Bertrand Editora; 2022

[29] Rajagopal KR. A note on the classification of anisotropy of bodies defined by implicit constitutive relations. Mechanics Research Communications. 2015;**64**:38-41

[30] Che SM, Spearman M, Manizade A. Constructive disequilibrium: Cognitive and emotional development through dissonant experiences in less familiar destinations. In: The Handbook of Practice and Research in Study Abroad. New York, USA: Routledge; 2010. pp. 121-138

[31] Fu K, Murphy J, Yang M, Otto K, Jensen D, Wood K. Design-by-analogy: Experimental evaluation of a functional analogy search methodology for concept generation improvement. Research in Engineering Design. 2015;**26**:77-95

[32] Alerasoul SA, Afeltra G, Hakala H, Minelli E, Strozzi F. Organisational learning, learning organisation, and learning orientation: An integrative review and framework. Human Resource Management Review. 2022;**32**(3):100854

[33] Christensen BT, Ball LJ. Creative analogy use in a heterogeneous design team: The pervasive role of background domain knowledge. Design Studies. 2016;**46**:38-58

[34] Easton GS, Rosenzweig ED. The role of experience in six sigma project success:

An empirical analysis of improvement projects. Journal of Operations Management. 2012;**30**(7-8):481-493

[35] Taylor A, Greve HR. Superman or the fantastic four? Knowledge combination and experience in innovative teams. Academy of Management Journal. 2006;**49**(4):723-740

[36] Milman O. The Insect Crisis: The Fall of the Tiny Empires That Run the World. Washington, D.C., USA: National Geographic Books; 2023

[37] Halliday T. Otherlands: A World in the Making-A Sunday Times bestseller. Penguin, England; 2022

[38] Schilthuizen M. Darwin Comes to Town: How the Urban Jungle Drives Evolution. London, England: Picador; 2019

[39] Mancuso S. La pianta del mondo. Gius: Laterza & Figli Spa; 2020

[40] Dennett DC. From Bacteria to Bach and Back: The Evolution of Minds. WW Norton & Company; 2017

[41] Rifkin J. The Zero Marginal Cost Society: The Internet of Things, the Collaborative Commons, and the Eclipse of Capitalism. St. Martin's Press; 2014

[42] Lipovetsky G. Le sacre de l'authenticité. Paris, France: Gallimard; 2021

[43] Kurniawan R, Manurung AH, Hamsal M, Kosasih W. Orchestrating internal and external resources to achieve agility and performance: The centrality of market orientation. Benchmarking: An International Journal. 2021;**28**(2):517-555

[44] Mutanen UM. Developing organisational design capability in a Finland-based engineering corporation: The case of Metso. Design Studies. 2008;**29**(5):500-520

[45] Bertola P, Teixeira JC. Design as a knowledge agent: How design as a knowledge process is embedded into organizations to foster innovation. Design Studies. 2003;**24**(2):181-194

[46] Rhee J, Park T, Lee DH. Drivers of innovativeness and performance for innovative SMEs in South Korea: Mediation of learning orientation. Technovation. 2010;**30**(1):65-75

[47] Buehring J, Bishop PC. Foresight and design: New support for strategic decision making. She Ji: The Journal of Design, Economics, and Innovation. 2020;**6**(3):408-432

[48] Wan CK, Lin SY. Negotiating social value, time perspective, and development space in sustainable product design: A dialectics perspective. Design Studies. 2022;**81**:101121

[49] Crilly N, Cardoso C. Where next for research on fixation, inspiration and creativity in design? Design Studies. 2017;**50**:1-38

[50] Dane E. Reconsidering the trade-off between expertise and flexibility: A cognitive entrenchment perspective. Academy of Management Review. 2010;**35**(4):579-603

[51] Nickerson RS. Confirmation bias: A ubiquitous phenomenon in many guises. Review of General Psychology. 1998;**2**(2):175-220

[52] Baer M, Brown G. Blind in one eye: How psychological ownership of ideas affects the types of suggestions people adopt. Organizational Behavior and Human Decision Processes. 2012;**118**(1):60-71

[53] Mahmoud NE, Kamel SM, Hamza TS. The relationship between tolerance of ambiguity and creativity in architectural design studio. Creativity Studies. 2020;**13**(1):179-198

[54] Stanley Budner NY. Intolerance of ambiguity as a personality variable. Journal of Personality. 1962;**30**(1):29-50

[55] Moreno DP, Hernandez AA, Yang MC, Otto KN, Hölttä-Otto K, Linsey JS, et al. Fundamental studies in design-by-analogy: A focus on domain-knowledge experts and applications to transactional design problems. Design Studies. 2014;**35**(3):232-272

[56] Eisenmann M, Grauberger P, Üreten S, Krause D, Matthiesen S. Design method validation–an investigation of the current practice in design research. Journal of Engineering Design. 2021;**32**(11):621-645

[57] Muratovski G. Paradigm shift: Report on the new role of design in business and society. She Ji: The Journal of Design, Economics, and Innovation. 2015;**1**(2):118-139